Wissenschaftliche Beiträge
zur Medizinelektronik

Band 8

Wissenschaftliche Beiträge zur Medizinelektronik

Band 8

Herausgegeben von
Prof. Dr. Wolfgang Krautschneider

Juan José Montero Rodríguez

Impedance spectroscopy for characterization of biological matter

Logos Verlag Berlin

λογος

Wissenschaftliche Beiträge zur Medizinelektronik

Herausgegeben von
Prof. Dr. Wolfgang Krautschneider

Technische Universität Hamburg-Harburg
Institut für Nano- und Medizinelektronik
Eißendorfer Str. 38
D-21073 Hamburg

Bibliografische Information der Deutschen Nationalbibliothek

Die Deutsche Nationalbibliothek verzeichnet diese Publikation in der Deutschen Nationalbibliografie; detaillierte bibliografische Daten sind im Internet über http://dnb.d-nb.de abrufbar.

ISBN 978-3-8325-4746-2
ISSN 2190-3905

Logos Verlag Berlin GmbH
Comeniushof, Gubener Str. 47,
10243 Berlin
Tel.: +49 (0)30 / 42 85 10 90
Fax: +49 (0)30 / 42 85 10 92
http://www.logos-verlag.de

Impedance spectroscopy for characterization of biological matter

Vom Promotionsausschuss der
Technischen Universität Hamburg

zur Erlangung des akademischen Grades

Doktor-Ingenieur (Dr.-Ing.)

genehmigte Dissertation

von

Juan José Montero Rodríguez

aus

San José, Costa Rica

2018

1. Gutachter: Prof. Dr.-Ing. Wolfgang Krautschneider
2. Gutachter: Prof. Dr.-Ing. Ralf Pörtner

Tag der mündlichen Prüfung: 15.08.2018

To my dear wife, Katherine.

Acknowledgements

I want to express my sincere gratitude to Prof. Dr.-Ing. Wolfgang Krautschneider for giving me the opportunity to pursue studies at the Institute, and for his continuous support and encouragement over the past four years. I also want to thank Dr.-Ing. Dietmar Schröder for his valuable feedback. Also I appreciate the help from all the academic and administrative staff of the Institute of Nano- and Medical Electronics. I thank Prof. Dr.-Ing. Ralf Pörtner for being the second reviewer of this dissertation, and together with Dr. Christiane Goepfert for the support with cell cultures at the Institute of Bioprocess and Biosystems Engineering.

I want to thank the academic staff from the Institute of High Frequency, especially Christian Friesicke, Carmen Hajunga, Malte Giese, Frauke Gellersen, Björn Deutschmann and Prof. Dr.-Ing. Arne Jacob for his support through the complete project. Also I express my gratitude to the academic staff from the Institute of Sports Medicine of the University of Hamburg, especially to Nils Schumacher and Mike Schmidt. Also I thank Kai Wellmann from Innovations- und Wissenstrategien GmbH, and Dr.-Ing. Stephan Brinke-Seiferth from i3membrane GmbH. I am also grateful to Prof. Dr. med. Udo Schumacher and Tobias Gosau from the Institute of Anatomy and Experimental Morphology, for providing me support with additional biological experiments at the University Medical Center Hamburg-Eppendorf.

I also thank my dear wife Katherine, for giving me support in the hard times, and for her sacrifice of accompanying me in this chapter of my life. I thank my parents Renato and Milagro, and my brother and sister, for always encouraging me to follow my goals. And to Dr.-Ing. Paola Vega-Castillo and Dipl.-Ing. Arnoldo Rojas-Coto, I am deeply grateful for their guidance and support in both academic and personal planes. I would not be standing here without their continuous support over the years.

I also want to thank the members of the TUHH SymphonING orchestra, and the conductor David Dieterle, for trusting me as a musician and for giving me countless hours of music, wonderful concerts, and the opportunity play in professional musical environments with the Göttlicher Philharmonie Hamburg and the Sinfonietta Nova.

Finally, I thank the Instituto Tecnológico de Costa Rica (ITCR) for funding of this research through the grant 34-2014-D, and for trusting me as part of the academic staff. I hope to contribute with this academic title to the prosperity of my country.

Abstract

This dissertation deals with electrical characterization methods of biological samples. Standard cell counting methods are time consuming and put stress on the cells by staining them with markers, which reduce cell viability. Therefore, there is much interest in alternative fast and non-destructive methods. Impedance spectroscopy is an electrical approach that gives information about its permittivity and conductivity of a sample. Each experiment produces an impedance 'fingerprint' which may be used to observe the evolution of a cell culture over time. Parameters such as the cell membrane capacitance or the conductivity of the cytoplasm affect the overall impedance, and are identified and extracted by mathematical modeling of the impedance data, using equivalent circuits based on electrochemical components such as resistors, capacitors, Warburg or constant-phase elements. Measurement over different frequencies is known to produce alpha, beta and gamma dispersions, which are regions where a specific physical effect takes place, such as the double-layer polarization, the opening or closing of an ionic channel in the membrane, or the water relaxation effect observed at microwave frequencies. Measurements are carried out with potentiostats, LCR meters or vector network analyzers, which are the industry standard for impedance characterization. The technique is used in the laboratory for identifying tumor cells in mice, comparing them with tissues from the liver, lung, brain, heart and leg. Other applications included in this work are the identification of bacteria in eye infections, the monitoring of the ripening process of fruits, and the detection of blood lactate concentration in athletes by using a skin sensor. These applications often require a portable measurement system. Therefore, a portable impedance spectroscopy system was designed and tested. A significant reduction of size, weight and power consumption of portable systems can be achieved by using an Application-Specific Integrated Circuit (ASIC). Three ASICs were used to assemble miniaturized devices. The ASIC 1 operates at frequencies below 40 kHz, and it is used for cell growth experiments. The ASIC 2 and ASIC 3 operate at frequencies close to 40 GHz and are suitable for impedance recording above the water relaxation frequency. It is concluded that the method works and can be effectively integrated into an ASIC for field experiments where portability is required. The method can be further improved by 4-terminal measurements. For extension of the method to millimeter-wave frequencies, full electromagnetic simulation of the chip has been carried out, and electrodes and interconnections have been adjusted accordingly.

Contents

Chapter 1

Introduction

The discovery of biological cells is attributed to Robert Hooke in 1665, who observed them for the first time on a bottle cork using a self-made microscope [1]. Cells are now considered the fundamental units of life, and it is known that all living organisms, plants and animals, are composed of these small units. It is estimated that the human body is composed of approximately 37.2×10^{12} cells [2], and each one of them is specialized to perform a specific function in the body, after a process known as cellular differentiation. This gives origin to more than four hundred types of cells in humans [3]. There are at least one hundred types of neurons, which are the cells responsible for the electric signaling and transmission of action potentials in the body. The large number of neuron types shows the importance of electricity in living organisms.

Cell cultures are used for multiple purposes, ranging from cancer research, virology and vaccine development, cytotoxic studies, genetic engineering, drug efficacy, and many others. To determine the number of viable cells in a culture, the cells are stained with markers, which makes them visible under the microscope, and they are counted optically to calculate the total concentration of cells. This process of manually counting the cells requires time and effort. Furthermore, cells may lose vitality when the staining markers are added. This is one of the reasons why there is much interest in alternative fast and non-destructive methods of cell counting. One of the possibilities is to use an electrical characterization method based on electrochemical impedance spectroscopy (EIS), which may give an indicator of the total number of viable cells in a culture. The method additionally gives information about the conductivity and permittivity of the cells over a wide range of frequencies, and this information could be used to identify biological samples, based on the impedance 'fingerprint' of each cell type.

Impedance is an electrical property describing the opposition to a current flow, and it is an indirect measurement of the permittivity and conductivity of a sample. The technical description of impedance and how to measure it using electronic devices is presented in Chapter 2. Early biological experiments compared the electrical conductivity of water to that of organic material, such as blood samples. For example, Höber measured in 1910 the conductivity of blood cells from pork and cattle, and compared them with ionic solutions [4]. This evolved into the field of bioelectrical impedance analysis (BIA), which is widely used nowadays in sports centers for measuring the total body water composition (TBW) and estimate body fat [5].

Impedance is typically measured with commercial devices such as potentiostats, galvanostats and LCR meters, which have a frequency limit in the MHz range. Vector network analyzers (VNA) may extend this range to GHz. The frequency is relevant because cells have a semipermeable membrane that behaves electrically as a capacitor, blocking the low-frequency electrical fields from penetrating the cell. Therefore, the use of high-frequency stimulation fields in the range of GHz could give additional information regarding the internal parts of the cells, since these fields can penetrate the cellular membrane.

The potentiostat was invented around 1940 by A. Hickling [6–9]. In the following years, electrochemical impedance spectroscopy (EIS) became an established industrial technique with applications in corrosion [10] and fuel cell characterization [11], and the fields of application of this technique started to emerge and diversify. Some of the known biological applications of impedance spectroscopy include cell growth monitoring and cell viability [12], bacteria detection and identification [13, 14], food quality control [15–18], cytotoxicity assessment [19], and indirect measurement of glucose [20], lactate [21] and other metabolites present in the body.

In 1967, Hewlett-Packard introduced the HP 8410 [22], one of the first high-frequency vector network analyzers (VNA) that expanded the frequency range up to 12 GHz. Nowadays, millimeter-wave network analyzers operate at frequencies above 100 GHz and beyond. With these advances, bioimpedance research has recently seen light in areas such as cancer identification using electrochemical impedance tomography (EIT) [23], with the demonstrated ability to distinguish between cancerous and healthy tissues for liver [24] in the frequency range from 0.5 to 20 GHz. This is one of the main motivations for the design of millimeter-wave ASICs for impedance spectroscopy.

1.1 Structure and classification of cells

Unicellular organisms such as bacteria or archaea are constituted by a single prokaryotic cell. In these type of cells, the cytoplasm contains a mixture of DNA, proteins and metabolites, without any additional organelle. These organisms are sized generally between 1-2 µm long and have a diameter of 0.25-1 µm. Some examples of prokaryotic organisms are the E. Coli and the Salmonella bacterias.

In contrast, multicellular organisms such as plants and animals are composed of eukaryotic cells. These have a membrane-delimited nucleus with the DNA, and separate compartments such as mitochondria, chloroplasts or vacuoles. They have diameters ranging from 5 to 100 µm. One example of eukaryotic cell is the S. cerevisiae, which is widely known for their use as baking yeast.

Cells may have different shapes depending on their function. For example, an erythrocyte or red blood cell is circular, but in contrast neurons have specialized structures such as axons and dendrites to propagate electrical impulses. Muscle cells such as myocytes are elongated, and contract to produce movement. Plant cells are often rectangular in shape and have a thicker cell wall composed of lipids and cellulose. They include additional structures such as chloroplasts, with the purpose of producing energy through photosynthesis.

The parts of plant and animal cells are summarized in Fig. 1.1.

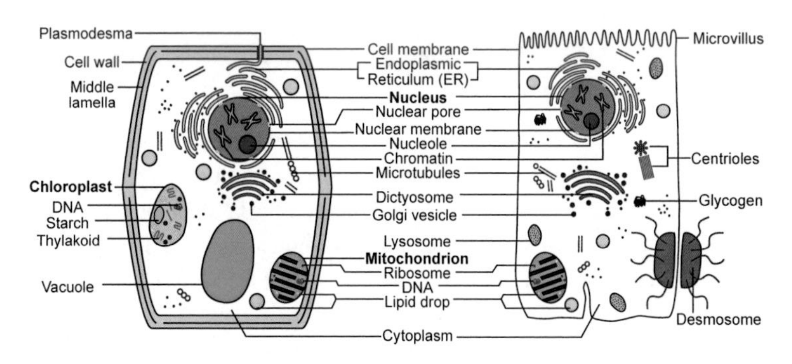

FIGURE 1.1: Parts of the plant and animal eukaryotic cells.

The cell membrane separates the contents of the cell from the environment. It is composed of a lipid bilayer which is hydrophobic, and allows some molecules to pass selectively during endocytosis and exocytosis. The nucleus contains the DNA which is required for cell reproduction, since it includes instructions to synthesize every protein, biomolecule and cell structure present in the cell. The DNA is packaged in the form of chromatin, which later constitutes the chromosomes. The endoplasmic reticulum together with the ribosomes allows the synthesis of proteins and lipids. Proteins are handed to the Golgi complex which sorts them into vesicles, addressed to be expelled from the cell or directed to other organelles. Lysosomes contain enzymes that break proteins, carbohydrates, lipids and nucleic acids, decomposing them into simpler molecules. These are found only in animal cells, as opposed to vacuoles, which perform similar functions on plant cells. Mitochondria and chloroplasts are the power source of the whole organism, since these organelles synthesize the energy molecules of ATP, the former through the aerobic process known as Krebs cycle, also known as the citric acid cycle, and the latter through the Calvin cycle during photosynthesis.

1.2 History of animal electricity

Animal electricity was first evidenced by Luigi Galvani in 1780 [25]. In one of his early experiments, he found that the muscles of a dead frog contracted and moved with the application of an electric discharge. The same movement was observed when lightning struck a wire connected to the animal's leg nerves. This was the first explanation of how nerves work in a biological organism: they are electrical conductors that carry signals to the muscles. The new discovery settled the foundations of electrophysiology, which is the study of the electrical properties of biological cells and tissues.

Galvani experimented with external sources of electricity, such as the charged metallic scalpel or an atmospheric discharge. Later he connected two different metallic electrodes and observed that the frog also twisted when the metals made contact, even when there was no apparent source of charge. He thought the frog was the source of the electrical charge, which was not totally mistaken. However, the experiments were repeated by Alessandro Volta, who discarded Galvani's theory because two different metals can generate an electric flow (a phenomenon now known as galvanic corrosion). Although Volta could reproduce Galvani's experiments, the source of bioelectricity in living organisms, without any external voltage or current application, was found later.

Another example of the presence of self-generated electric fields in the organism is the process of wound healing of the skin, observed by Du Bois-Reymond in 1848

[26]. When a wound is made in the skin, an electric field is immediately formed. This electric field guides cell migration towards the wounded area. This gave origin to recent wound healing techniques for muscular rehabitlitation, such as transcutaneal electrical nerve stimulation (TENS) [27].

It is now known that the cell membrane acts like an electrical capacitor, accumulating charge between the interior and exterior of the cells. Ions like sodium (Na^+) and potassium (K^+) enter the cell through ionic channels [28], and they are accumulated in equilibrium concentrations at the inside and outside of the cells. This gives origin to the *transmembrane potential*, a potential difference across the cell membrane, which in resting condition is typically between –40 mV and –80 mV with respect to the exterior of the cell. When cells are activated, this *resting potential* may suddenly reverse by *depolarization* and give origin to an *action potential* [29], which is propagated through the nerves to perform muscular movement. Afterwards, *repolarization* returns the cell potential to the resting value.

The activity of ionic channels, and therefore the electrical activity of the cells, is studied by the voltage clamp, the current clamp and the patch clamp techniques. These methods allow recording the electrical activity of one individual cell. The voltage clamp experiment is used to observe the ionic current flowing through the cellular membrane, by setting and holding the membrane voltage at a fixed level. This is achieved by using a negative feedback amplifier. The current clamp experiment injects a constant current into the cell, using one of the recording electrodes. The membrane voltage is recorded externally. This method has been used effectively to record the firing of a neuron when it is depolarized [30].

The patch clamp technique, developed by E. Neher [31] and B. Sakmann [32], allows recording ionic activity at a smaller portion of membrane. A micropipette is filled with an electrolyte solution and a recording electrode is placed inside. The micropipette is placed close to the cell, and suction is applied to grab a small portion of the membrane. If more suction is applied, the membrane breaks and the cytoplasm is combined with the electrolyte, forming a whole-cell clamp. Other variations of the method include pulling the membrane to form an inside-out or outside-out clamp. After this procedure, either the voltage clamp or current clamp experiments are performed. The method and variations illustrated in Fig. 1.2.

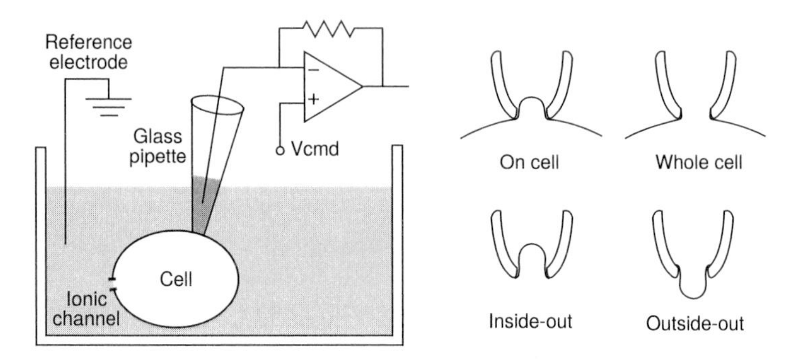

FIGURE 1.2: Patch clamp experiment and configurations [31].

1.3 Field exposure of cells and biological samples

The interaction of cells with artificially generated electric fields has been observed experimentally in a large collection of empirical studies. The stimulation of a cell culture with external electromagnetic fields may induce different effects on the cell physiology and morphology, depending on the amplitude and frequency of the voltages and currents applied. Cells can rotate, align or migrate towards the electrodes. A specific field may kill the cells or inhibit their ability to grow, whereas other frequencies or amplitudes may enhance cellular reproduction. These findings have been used for the development of cancer-treating therapies, and for regeneration therapies such as tissue stimulation.

Until now, the experimental results show the effects of field exposure on the growth rate and health state of the cells, but an explanation of the underlying mechanisms of action is still missing. It is known that cellular signaling occurs both in chemical and electrical manners, but it is not clear why certain frequencies or pulse widths accelerate the growth rate of a particular cell culture, and could be lethal for another type of cells.

When a direct-current (DC) field is applied, cells tend to rotate, align and migrate in the direction of the external field [33]. This phenomenon is known as galvanotaxis, with patented applications in cell separation using electric fields [34]. Direct-current fields have shown an increase in the speed of wound healing [35] and bone fracture regeneration [36]. These methods work by replicating the endogenous electric field produced in the organism when there is a wound, stimulating the organism to produce cells faster and directing them to the desired location. Wound healing has been addressed in a review of 119 animal and human experiments by Spadaro [37], where the majority of studies apply DC electric fields, and pulsed DC fields in the frequency range from 0.5 to 500 Hz.

Alternating-current (AC) has shown effects on the growth speed of cell cultures when the stimulation is performed at low frequencies. For example, Rivera [38] observed marked growth acceleration on vaginal wall muscular cells, after three days of stimulation at 200 Hz when compared to a control cell culture. The same growth acceleration effect was observed by Chang [39] on fetal neural stem cells under stimulation at 100 Hz.

Electroporation is the ability to induce pores or channels in the membrane, with the purpose of introducing substances that otherwise cannot pass through the membrane. For example, the introduction of DNA into a cell is called transfection, and this is described in a protocol by Covello [40]. Transfection has applications on gene expression and targeted gene silencing. Commercial devices include the Neon MPK5000 from Invitrogen [41], or the Gene Pulser Xcell System from Bio-Rad [42]. These devices work by applying high-voltage pulsed DC fields with short pulse widths. The amplitude of the stimulation voltage varies from 10 V to 3000 V with exponential or square waveforms.

Exposure to specific frequencies and pulse widths may also induce cell death by apoptosis, which has applications on cancer treatments. These particular fields are often referred as tumor-treating fields (TTF) [43] and nanosecond pulsed electric fields (nsPEF). The tumor therapy by TTF works by breaking the highly dividing cells at specific orientation [44], and it is selective because it only kills cells undergoing cellular division. In the latest part of the cellular reproduction cycle, known as cytokinesis, the cell forms a channel which can be broken prematurely by induced electric fields. With the appropriate intensity and orientation, the cellular membrane breaks, killing the dividing cell before it finishes the reproduction cycle. Commercial devices such as the Novo-TTF operate at frequencies between 100-300 kHz, citing the optimal frequencies for different types of cells as “100 kHz for mouse melanoma (B16F1), 150 kHz for human breast carcinoma (MDA-MB-231), and 200 kHz for rat glioma (F-98)” [43].

1.4 Purpose of the work

Electronics engineering has shown significant advances in the last seventy years. The first transistor was built in the Bell laboratories around 1947, and nowadays a modern microprocessor features more than one billion of MOS transistor devices, with the number doubling roughly every two years as predicted by the empirical Moore's Law, published in 1965 [45]. Miniaturization of electronic components opened new possibilities in medical electronics, with the integration of advanced devices on an Application-Specific Integrated Circuit (ASIC). This gives the possibility of implanting electronic devices in the body, with the challenges being low power consumption and wireless data communication. For example, implants may be used to monitor parameters such as pressure and temperature inside the body, and this may help doctors to prevent and diagnose problems ahead of time.

This dissertation addresses the medical and biological applications of impedance spectroscopy, by using commercial measurement equipment, electronic circuits, and ASICs for field exposure and impedance spectroscopy characterization. The technique is used to differentiate between liquid samples infected with a bacteria, as opposed to samples from healthy patients at the University Medical Center Hamburg-Eppendorf (UKE). Another application is the monitoring of the ripening process in strawberries, which may be useful in the future to observe and compare the effect of preservatives added to the food. The next application is an exploratory study to identify tumor tissue in mice, comparing the impedance of tumor cells to that of healthy tissues from the heart, brain, lungs or leg muscles. The last application included in the dissertation includes a preliminary study for the development of a sports sensor, using a special electrode configuration, with the purpose of monitoring the lactate concentration in blood. This study was performed in cooperation with the Institute of Sports Medicine from the Hamburg University and the company i3membrane GmbH.

Miniaturization of electronic components also gives the possibility of developing portable electronic devices for impedance monitoring in environments requiring mobility, such as an athlete running a marathon, where these type of devices could monitor physiological quantities such as the lactate concentration in sweat. This is one of the main motivations of this work: the use of electronic devices to further miniaturize impedance spectrometers for portable biomedical applications. In the sports example, an impedance monitoring device could fit in an armband or smart watch, and provide instant feedback to the athlete during intensive training.

In this direction, three ASICs have been previously designed at the Institute of Nano- and Medicine Electronics of the Hamburg University of Technology, and they are used in this dissertation to assemble miniaturized impedance monitoring devices. The ASIC 1 is implemented in the AMS H35B4 350 µm technology, and measures at frequencies up to 40 kHz [46]. The ASIC 2 and ASIC 3 are designed in the IHP SG13S 130 µm technology, and operate at frequencies between 40 and 50 GHz. A photograph of each device is shown in Figure 1.3.

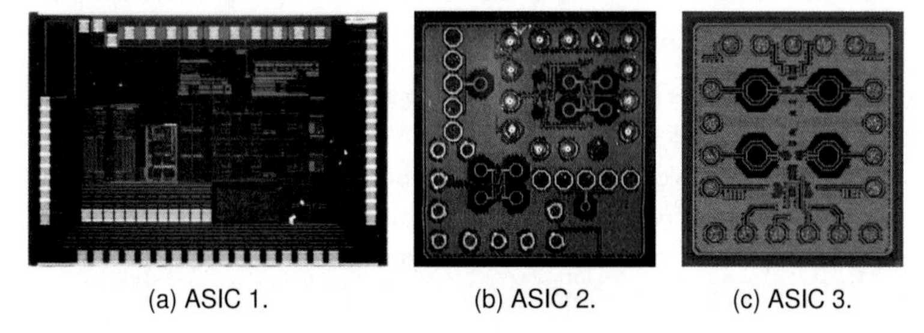

(a) ASIC 1. (b) ASIC 2. (c) ASIC 3.

FIGURE 1.3: ASICs designed at the Institute of Nano- and Medicine Electronics.

These devices are considerably smaller than a bench potentiostat or a commercial LCR meter, they consume less power and may be integrated into standalone portable devices, without the need of a computer for data recording. As a comparison, a standard LCR meter could weight several kilograms, use power from the 230 V power grid, and measure about 40x40x20 cm. The integrated circuits by contrast weight a few grams, operate on reduced voltages of 3.3 V or 1.2 V, and measure less than 1x1 mm. These devices apply a small electric field to the cells, and record the impedance during stimulation. The ASICs are used to perform biological experiments with cell cultures under different environmental conditions, such as the presence of sugar in the culture media, which accelerates the growth rate as compared to culture media without sugar. This was observed in the impedance, showing the contrast between the two experiments.

The design and assembly of test fixtures, and finally, miniaturized devices for biological experiments with the ASICs is challenging due to the small physical dimensions of the devices. As an example, the pad size for ASIC 3 is of 55 µm diameter, and the pad separation is of 130 µm, making it difficult to connect to external devices. Additionally, the last two ASICs operate in millimeter-wave frequencies, and any external connector or wire introduces undesired parasitic effects.

1.5 Outline

Chapter two includes the theoretical background for impedance spectroscopy of cell cultures, the two-, three- and four-electrode configurations, followed by the modeling process of spectroscopy data with equivalent circuit components. A simulation of a cell culture with COMSOL® is proposed based on the dielectric mixture theory. The chapter concludes with the development of a toolbox for fitting impedance data to circuit models using the nonlinear least-squares method in MATLAB®.

Chapter three includes a collection of experiments in different applications related to electrochemical impedance spectroscopy, such as measurements for fruit ripening, cancer detection in mice, the identification of bacterial infections in liquid samples from the human eye, and experiments for lactate measurements in sweat and blood. These were carried out using standard industrial equipment and proved that the method works for a variety of applications.

The development of a portable impedance spectroscopy device requires a review of the circuitry of such devices. Chapter four reviews common circuit topologies of impedance measurement devices, and how they are implemented using electronic components, together with the relevant calculations for obtaining the impedance curves based on the electrical measurements.

Chapter five includes the design and implementation of a portable impedance measurement system using commercial components, by using the Auto-Balancing Bridge Method (ABBM). A demonstrator of this device is working at 125 kHz and is used to monitor the impedance of yeast cell cultures. The device exhibits an accuracy level of up to 95% when compared with an LCR meter. This work was published in [47].

Further miniaturization of the impedance measurement device involves the use of an Application-Specific Integrated Circuit (ASIC). Chapter six deals with the development of a test board in the low frequency range, under 40 kHz, using the ASIC 1. This device is used to record the impedance of yeast and porcine chondrocytes, showing impedance variations as the cells grow under the presence or absence of sugar. These experiments were published in [46].

An extension of the method to the high-frequency range could give additional dielectric information regarding the inner parts of the cell, such as the nuclear envelope and organelles, at frequencies above the water relaxation effect. Chapter seven contains the design specifications, simulation and characterization of the high frequency devices up to 50 GHz, focusing on the ASIC 2. This includes measurements of the DC operating point, S-parameters of the integrated inductors and the VCO oscillation frequency.

Chapter eight includes the design specifications, simulation and characterization of the ASIC 3, and the development of test fixtures using flip-chip and wire bonding. The DC operating point and VCO oscillation frequency are measured. A custom test board was fabricated using two stacked Printed Circuit Boards (PCB). The bottom PCB includes a Rogers 4003 substrate for power connections, and the top PCB consists of an aluminium oxide substrate with a microstrip line, used as a sensor for cell characterization. This test board is used to perform cell experiments with CHO DP-12 cells at 43 GHz. The chapter ends with a full electromagnetic simulation of the assembly.

Chapter nine is the conclusion and outlook of the dissertation.

Chapter 2

Impedance spectroscopy

Electrochemical Impedance Spectroscopy (EIS) measures the conductivity and permittivity of a sample as function of the frequency. The analysis and modeling of impedance data can give information regarding the interfaces between different metals and electrolytes in the experimental setup.

The sample is stimulated by an alternating current, and the voltage is measured simultaneously to determine the impedance. Impedance is calculated by means of the Ohm law, and it is a complex variable, depending on the phase angle of the voltage and the current. The real part of the impedance is proportional to the resistivity, and the imaginary part corresponds to the reactance.

The impedance is calculated using the following equations, in the sinusoidal, Euler and phasor representations:

$$Z(\omega) = \frac{v(\omega)}{i(\omega)} = \frac{v_0 \sin(\omega t + \theta)}{i_0 \sin(\omega t)} = \frac{v_0 \cdot e^{j(\omega t + \theta)}}{i_0 \cdot e^{j\omega t}} = |Z| \cdot e^{j\theta} = \frac{v_0 \angle\theta}{i_0 \angle 0^\circ} = |Z| \angle\theta \qquad (2.1)$$

These equations consider v_0 and i_0 to be small to assume linearity in the system (small-signal AC analysis). The voltage applied to the sample may optionally have a DC offset which does not affect the small-signal measurement, although it may affect the physical system as it was discussed in Section 1.3.

Data from an EIS experiment can be plotted in several formats. As an example, the impedance of a concentrated aqueous solution of calcium chloride ($CaCl_2$) was measured with a Gamry Interface 1000E^{TM} potentiostat using a chamber of 1 mm^3 with two stainless-steel electrodes. The sample is diluted by half and the experiment is repeated several times. Each curve in the diagrams represents a different ionic concentration.

The polar Bode plot of Fig. 2.1 includes the magnitude and phase of the impedance as a function of frequency. The rectangular Bode plot is shown in Fig. 2.2, with the real and imaginary parts of the impedance. The Nyquist plot shows the real part in the horizontal axis, and the imaginary part in the vertical axis, as shown in the left side of Fig. 2.3. The 3D representation is shown in Fig. 2.3.

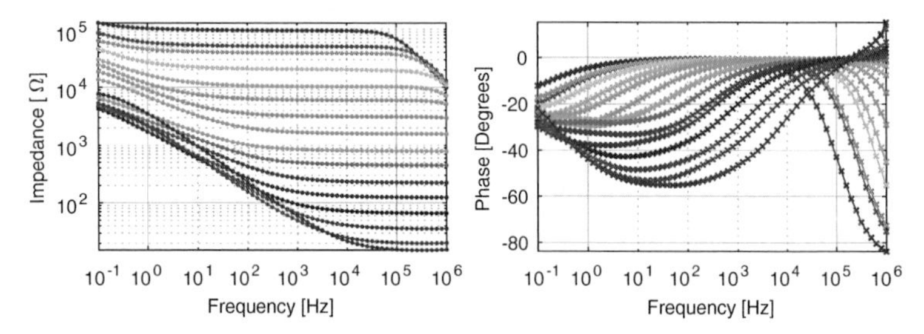

FIGURE 2.1: Polar Bode plot showing the magnitude (left) and phase (right) of the impedance as a function of frequency.

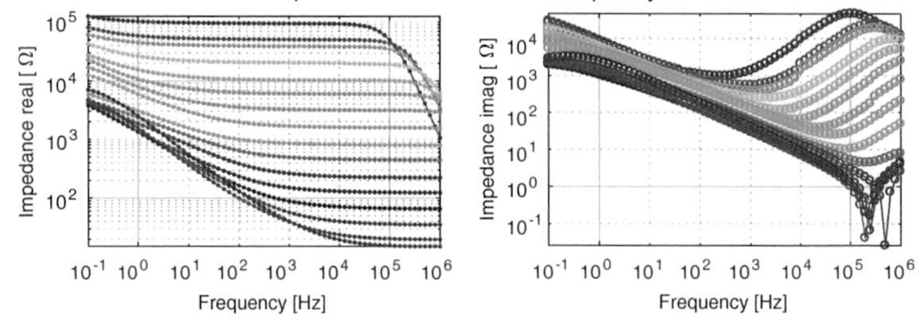

FIGURE 2.2: Rectangular Bode plot showing the real (left) and imaginary (right) parts of the impedance as a function of frequency.

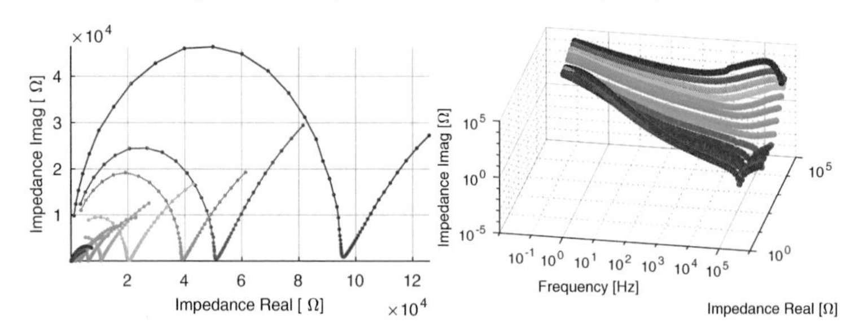

FIGURE 2.3: Nyquist plot (left) and 3D representation (right) of the impedance.

The impedance contains information related to the conductivity and permittivity of the analyzed sample. These physical constants are calculated by means of the resistance and reactance equations, as shown in the following section.

2.1 Impedance, permittivity and conductivity

Impedance is related directly to the permittivity and conductivity by the following equations:

$$Z(\omega) = R(\omega) + jX(\omega) \tag{2.2}$$

with R being the resistance, and X the reactance or complex impedance.

Assuming a parallel-plate chamber with two electrodes, with surface area A, and separated by a distance d. The definitions of resistance ($R = \rho l/A$) and capacitance ($C = \epsilon A/d$) are substituted as follows:

$$Z(\omega) = \frac{\rho(\omega)d}{A} + \frac{1}{j\omega C(\omega)} \tag{2.3}$$

$$Z(\omega) = \frac{\rho(\omega)d}{A} + \frac{1}{j\omega} \cdot \frac{d}{\epsilon(\omega)A} \tag{2.4}$$

As shown in the previous equation, the resistivity $\rho(\omega)$ and permittivity $\epsilon(\omega)$ are frequency-dependent parameters, and they are calculated using the geometrical parameters of the experimental setup. These parameters are often grouped as a single geometric constant G, as follows:

$$Z(\omega) = \frac{d}{A}\left(\rho(\omega) + \frac{1}{j\omega\epsilon(\omega)}\right) \tag{2.5}$$

$$Z(\omega) = G\left(\frac{1}{\sigma(\omega)} + \frac{1}{j\omega\epsilon(\omega)}\right) \tag{2.6}$$

This constant can be calculated from the previous equations, or it can be measured empirically using a standard conductivity solution. If the experimental setup is more complicated, such as a four-electrode experiment, the distance between the electrodes is not clearly defined, and therefore this constant cannot be calculated analytically. However, an effective geometrical constant may still be determined empirically by measuring the impedance of a standard conductivity solution.

2.2 Cellular impedance recording

The permittivity and conductivity of biological tissues and cell cultures change in steps depending on the measurement frequency, as shown in Figure 2.4 [48]. Each step on the diagram is known as a dispersion region, and is caused by a specific set of physical effects. There are four identified dispersion regions known as alpha, beta, gamma and delta. A detailed review of the dispersion regions can be consulted in [49], summarized briefly in the following paragraphs.

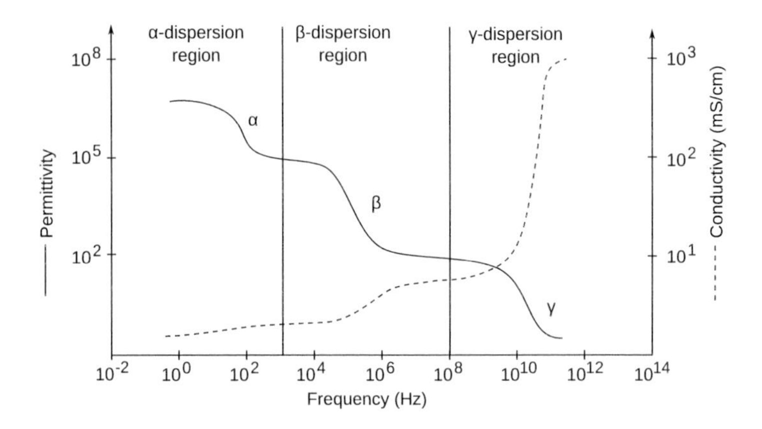

FIGURE 2.4: Frequency dispersion regions for cells [48].

The alpha dispersion (1 Hz up to 1 kHz) is caused by the double-layer effect and the slow movement of the ions at the proximity of the electrodes. In addition to the double-layer effect, the alpha dispersion also depends on the ion channels of the membrane, described by the Hodgkin-Huxley model.

The beta dispersion (1 kHz to 100 MHz) occurs due to the membrane capacitance, which has an approximate value of 1 $\mu F/cm^2$, constant for most types of cells [50]. The membrane capacitance blocks low-frequency electric fields from penetrating the cells. As the frequency increases, this capacitance is short-circuited, allowing the electric fields to enter the cells and giving origin to the beta relaxation.

The gamma dispersion (above 100 MHz) is explained by the bound water and water relaxation effects. Water is a polar molecule, and in equilibrium conditions (without external fields) the dipoles align with themselves at regions known as bound water. External electric fields change the orientation of the dipoles and break bound water, causing a first dispersion. Above 10 GHz the dipoles cannot follow the field anymore, and the water relaxation effect is observed.

The delta dispersion is found at even higher frequencies and could give information about the internal elements of the cells, such as organelles, proteins, aminoacids and other molecules present in the interior of the cell membrane. These effects may be observed in the microwave spectrum (roughly between 300 MHz and 300 GHz).

The dispersion regions may be expressed by the Debye relaxation equation:

$$\epsilon(\omega) = \epsilon_\infty + \frac{\Delta\epsilon_1}{1 + i\omega\tau_1} + \frac{\Delta\epsilon_2}{1 + i\omega\tau_2} + ... \tag{2.7}$$

where ϵ_∞ is the permittivity at infinite frequency, $\Delta\epsilon_n$ is the permittivity change caused by the nth dispersion region, and τ_n is the relaxation time.

According to Schwan [48], the individual components of the cells that may appear in a permittivity dispersion graph are shown in Table 2.1. The presence of a dispersion is marked with a bulled.

TABLE 2.1: Dispersion regions for different tissues [48].

Biological material element		**Dispersion**			
		α	β	γ	δ
Water and electrolytes				•	
Macromolecules	Aminoacids			•	•
	Proteins		•	•	•
	Nucleic acids	•	•	•	•
Vesicles	Surface charged	•	•	•	
	Non-surface charged		•	•	
Cell membrane	+ Fluids free of protein		•	•	
	+ Proteins		•	•	•
	+ Surface charged	•	•	•	
	+ Membrane relaxation	•	•	•	
	+ Organelles		•	•	•
	+ Tubular system	•	•	•	
Cell with membrane, surface charge, organelles, proteins		•	•	•	•

The table shows the potentials of impedance spectroscopy for analyzing biological samples in different frequency regions.

2.3 Electrode configurations for EIS

Impedance spectroscopy can be performed industrially with an LCR meter, a potentiostat or galvanostat, or a Vector Network Analyzer (VNA). Each of these devices can be calibrated to reduce parasitics due to the connectors and wiring.

A standard potentiostat has four electrodes: the Counter Electrode (CE), Working Electrode (WE), Reference Electrode (RE), and Working Sense Electrode (WSE).

The working electrode is the voltage source of the device, and provides current to the sample. The counter electrode is the return path to ground, and it is often used to measure the current flowing through the cell. The reference and the working sense electrodes are used to measure the potential drop across the cell, after the double-layer polarization effect, in a region of the electrolyte where the electric field is uniform.

The electrodes and their connections with the internal components of an impedance measurement device are shown in Figure 2.5. If the experiment is potentiostatic, the voltage at the counter electrode is constant, whereas for a galvanostatic experiment, the current at this electrode is fixed.

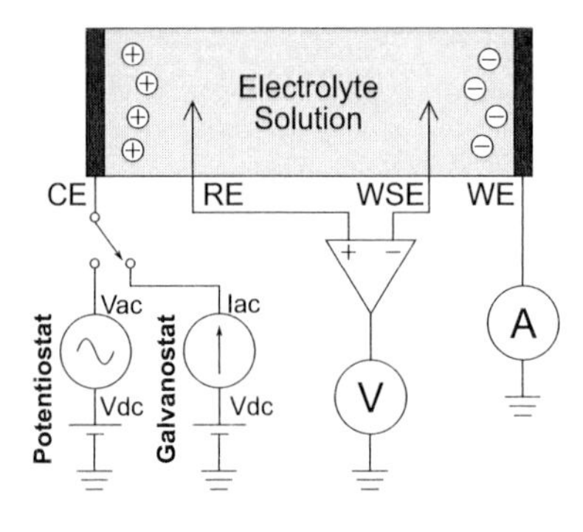

FIGURE 2.5: Diagram of a potentiostat/galvanostat and its four electrodes.

The counter and working electrodes (CE and WE) are often interchangeable, since they provide a closed current path to the sample. However, care should be taken when placing the voltage sampling electrodes (RE and WSE), because a wrong sensing polarity may result in an automatic increase of the stimulation voltage.

The four electrodes can be connected to the experimental chamber in three different configurations. These are described in Fig. 2.6 and explained as follows.

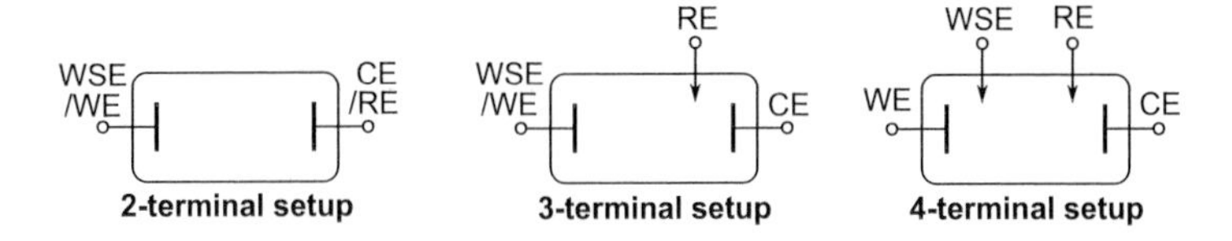

FIGURE 2.6: Electrode configurations for impedance spectroscopy.

2.3.1 Two-electrode configuration

The stimulation field is applied and the voltage is measured using the same pair of electrodes. This configuration is the simplest, and the experimental chamber can be designed using parallel electrodes. However, this has the disadvantage of being sensitive to the double-layer effect, which introduces a large parasitic impedance at frequencies close to DC.

2.3.2 Three-electrode configuration

In this configuration, the stimulation signal is applied between the CE and WE, and the reference potential is measured from the experiment through the RE, by using a porous electrode placed in contact with the aqueous solution medium, in a section of the chamber far away from the electrodes where the potential is known to be stable.

2.3.3 Four-electrode configuration

As in the previous two configurations, the stimulation signal is applied between the CE and WE. The other two electrodes, RE and WSE, are used to sample the voltage at a different position, apart from the electrodes, in a section of the sample where the electric field is assumed to be uniform. This allows to reduce the screening effect of the double-layer polarization at the electrodes, and measure the dielectric properties of the bulk electrolyte itself with better accuracy.

2.3.4 Accuracy contour plots

Accuracy contour plots describe the frequency and impedance range where the measurement can be assumed to be accurate, down to a specific confidence level. A typical accuracy contour plot for a generic potentiostat is shown in Fig. 2.7.

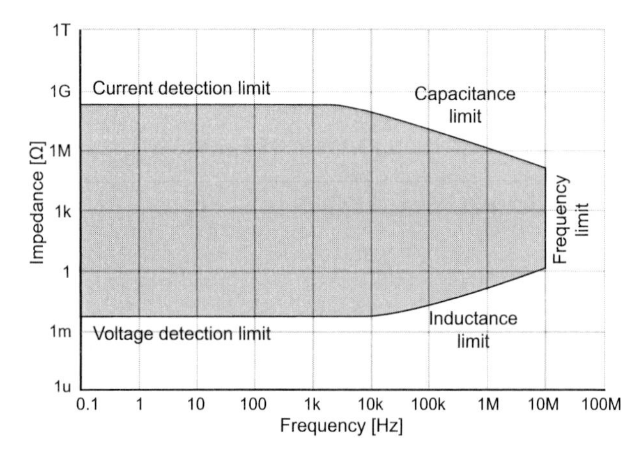

FIGURE 2.7: Accuracy contour plot for an impedance measurement device.

The limits of a potentiostat are determined by a short-circuit and open-circuit measurement. In the short-circuit test, the smallest measurable impedance is limited by the inductance of the wires and the smallest ADC resolution for voltage detection. In the open-circuit test, the largest measurable impedance is limited by the parasitic capacitance and the resolution for the smallest detectable current.

2.4 Equivalent circuit components

Impedance data is often fitted to a circuit model for a detailed analysis of the electrochemical reactions, charge accumulation, solid-liquid interfaces, double-layer formation and other physical effects. Models are selected considering parameters such as the electrode area, or the presence of reduction-oxidation reactions at the electrode surface. A list of commonly used equivalent circuit elements is presented in Table 2.2, with the corresponding impedance expressed as a function of the frequency and the phase angle.

TABLE 2.2: Equivalent circuit elements used for electrochemistry.

Element	Name	Equation	Phase
R	Resistor	$Z = R$	$0°$
L	Inductor	$Z = j\omega L$	$+90°$
C	Capacitor	$Z = 1/(j\omega C)$	$-90°$
W	Warburg Element	$Z = A/(j\omega)^{0.5}$	$-45°$
CPE	Constant-Phase Element	$Z = A/(j\omega)^{\beta}$	$-90° \times \beta$

These equivalent circuit components with their equations are used later to fit experimental data, for instance from sports lactate experiments in Chapter 3, and this modeling process allows to extract the resistivity of sweat. The elements from Table 2.2 are explained below.

Other notable effects observed in general electrochemical systems are the faradaic and non-faradaic charge-transfer processes, and the double-layer effect of electrode polarization. These physical effects are modeled by equivalent circuits, and they are described also in the following sections.

2.4.1 Resistance, capacitance and inductance

The resistor, capacitor and inductor are common electrical elements, and they are used in the same sense for electrochemical models. A resistor is an opposition to the flow of charge. Capacitance models charge accumulation at most of the electrode-electrolyte interfaces. Inductance does not usually appear as a direct electrochemical effect, but inductors are used to complement the models with the parasitics of wiring from the instrument to the experimental chamber.

The charge-transfer resistance (R_{ct}) is an element found often in electrochemical interfaces. This resistance models the transfer of charge from a solid electrode to a liquid electrolyte. At equilibrium conditions, a constant current density J is maintained due to faradaic reactions involving reduction and oxidation processes. The expression for the charge-transfer resistance is the following:

$$R_{ct} = \frac{RT}{zFJ} \tag{2.8}$$

where R is the gas constant, T is the temperature in Kelvin, z is the number of electrons transferred, F is the Faraday constant and J is the current density.

2.4.2 The Warburg element

The Warburg element is used to model diffusion processes of charged ionic species in the electrolyte. This element has a constant phase angle of –45°, and since the real and imaginary parts are identical, this results in a straight line on the Nyquist plot. The generalized Warburg equation considers semi-infinite linear diffusion, this means the ions have no limitations for the distance they can travel, and the equation is given as:

$$Z_W = \frac{A_W}{\sqrt{\omega}} + \frac{A_W}{j\sqrt{\omega}} = \frac{A_W}{\sqrt{\omega}}(1-j) = \frac{A}{\sqrt{j\omega}} \tag{2.9}$$

where A_W is the Warburg coefficient, with units of $[\Omega \cdot s^{-1/2}]$.

The Warburg coefficient may be calculated as:

$$A_W = \frac{RT}{z^2F^2A_e\sqrt{2}}\left(\frac{1}{C^o\sqrt{D_o}} + \frac{1}{C^r\sqrt{D_r}}\right) \tag{2.10}$$

where R is the ideal gas constant, T is the temperature in Kelvin, z is the number of electrons transferred, F is the Faraday constant, A_e is the electrode surface area, C^o and C^r are the concentration of oxidant and reductant species in the proximity of the electrode, and D_o and D_r are the diffusion coefficients of the oxidant and reductant.

2.4.3 The constant-phase element

The constant-phase element (CPE) is a generalization of the Warburg element. The CPE also produces a straight line in the Nyquist plot. The difference is the phase, which can be fixed to an arbitrary value depending on the exponent of the frequency term. The equation for this component is:

$$Z_{CPE} = \frac{A_W}{(j\omega)^\beta} \tag{2.11}$$

The phase of this element is given by the exponent β. For an exponent of β=0 this element is a resistor with A_W=R. If the exponent is β=1 this element describes a capacitor with A_W=1/C. An exponent of β=0.5 results in a Warburg element with a phase of 45°. The exact phase is calculated in degrees as $\phi = -90^\circ \times \beta$.

2.4.4 Faradaic and non-faradaic processes

When an electric signal is applied from a metallic electrode to an aqueous solution such as a cell suspension, the electrons, which are the electrical charge carriers, encounter a physical barrier. In this electrode/electrolyte interface, there are two main types of charge transfer processes: faradaic and non-faradaic [51].

Faradaic processes occur when charge is transferred directly across the interface. This electron transfer causes oxidation or reduction reactions, which cause a mass accumulation or loss at the electrode, explained by Faraday's Law of electrolysis. This equation is useful, for example, for controlling the electrodeposition of conductive polymers (such as PEDOT) over metallic electrodes. If a constant current is flowing through the interface, the mass that is deposited or lost in this electrochemical reaction is given by:

$$m = \frac{I \cdot t}{F} \cdot \frac{M}{z} \tag{2.12}$$

where m is the mass, I is the current, t is the total time of the current application, F=96485 C/mol is the Faraday constant, M is the molar mass of the material deposited and z is the number of electrons exchanged per molecule.

In non-faradaic processes, electrons cannot flow across the interface. Instead, they are accumulated in the metal, producing electrode polarization. The ions close to the surface of the polarized electrode will be attracted or repelled depending on their own charge, and an ionic drift current is induced in the electrolyte, without electron transfer or electrochemical reactions.

Both faradaic and non-faradaic processes may occur simultaneously in a given electrode-electrolyte interface, allowing charge transfer by both electrochemical reactions and the drift of ions due to electrode polarization. In addition, diffusion of ions will also be present in the bulk electrolyte. The equilibrium condition determines how much charge is allowed to flow.

Electrode polarization causes an additional effect known as the double-layer capacitance, which is explained as follows.

2.4.5 Double-layer capacitance

When electrode polarization occurs, the charged ions present in the electrolyte are pulled towards the electrode surface. In the Helmholtz model [52], specifically adsorbed ions form a thin layer known as the inner Helmholtz plane (IHP). These ions are bound by short-range interactions to the electrode. A second layer of ions of opposed charge to the electrode polarity appears in the electrolyte, in a region known as the outer Helmholtz plane (OHP). The Guoy-Chapman model then describes the diffuse layer, formed immediately after, with an ionic concentration gradient extending to the bulk solution. This arrangement stores charge in the double-layer, acting as a capacitor.

A graphical representation of the double-layer effect is shown in Fig. 2.8.

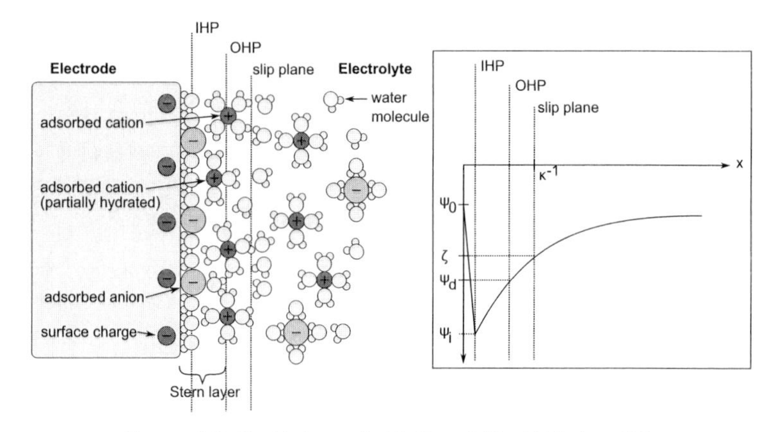

FIGURE 2.8: Double-layer effect in the solid/liquid interface [53].

In this drawing, the ζ potential indicates the point of zero charge, creating an equipotential line known as the slip plane. This region delimits the double layer. At the left of the slip plane, the ions are stuck to the surface. At the right side, ions are allowed to move freely and the ionic concentration gradient is observed.

The thickness of the complete double-layer (from the electrode to the slip plane) is given by the Debye length κ^{-1}:

$$\kappa^{-1} = \sqrt{\frac{\epsilon_r \epsilon_0 k_B T}{2 N_A e^2 I}} \tag{2.13}$$

where I is the ionic strength of the electrolyte in mol/L, ϵ_r is the dielectric constant, ϵ_0 is the permittivity of free space, k_B is the Boltzmann constant, T is the temperature in Kelvin, N_A is the Avogadro number and e is the elementary charge.

2.5 Equivalent circuit models

The following circuits are commonly used for fitting electrochemical processes. These circuits have a specific frequency response, describing different shapes on the Nyquist diagram. More information can be found at [54].

2.5.1 Simple electrode

The simplest model of an electrode-electrolyte interface consists of a solution resistance R_s, in series with the double-layer capacitance C_{dl}. The first element describes the resistivity of the liquid solution, and the second element models the ionic accumulation at low frequencies close to DC, describing the electrode polarization. The circuit element and the simulated response is shown in Fig. 2.9.

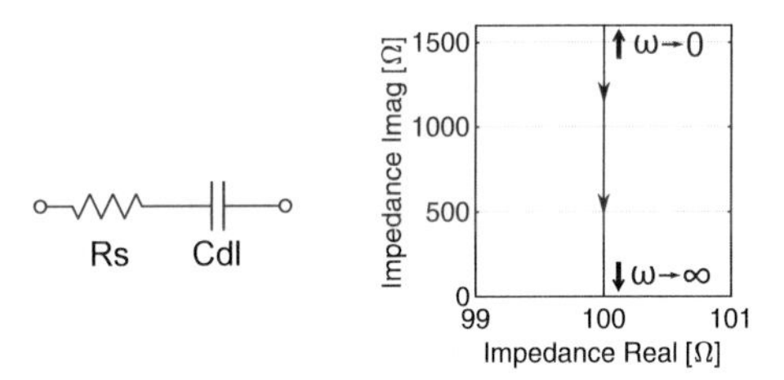

FIGURE 2.9: The simple electrode model and its simulated frequency response, with R=100Ω and C=1µF. The arrows indicate an increase in frequency.

The time constant of the circuit is $\tau = R_s C_{dl}$ and the cut-off frequency is given by $f_c = 1/2\pi R_s C_{dl}$. Below this frequency, the impedance increases as the frequency decreases until no charge flows at DC, similar to an open-circuit of a fully charged capacitor. Above the cut-off frequency, the capacitive reactance becomes insignificant. At f=∞, Z=R_s. The response of this circuit is described by the equation:

$$Z = R_s + \frac{1}{j\omega C_{dl}} \tag{2.14}$$

2.5.2 Randles cell

This model was proposed by Randles [55] to describe electrochemical reactions occurring at the electrodes. The model consists of a solution resistance (Rs), a charge-transfer resistance (Rct) and the double-layer capacitor (Cdl). Additional components such as the Warburg element are included to describe the diffusion of ionic species in the aqueous media.

The Randles circuit and variations including a CPE or a Warburg element are shown in Fig. 2.10. The three models describe a single frequency relaxation, observed in the Nyquist plot as a semicircle. The difference is the tail at the right side of the figures, corresponding with the lower frequencies of the measurement.

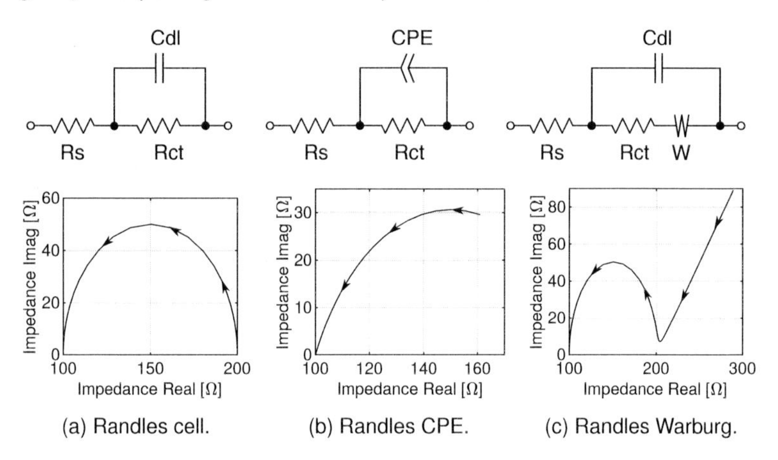

(a) Randles cell. (b) Randles CPE. (c) Randles Warburg.

FIGURE 2.10: Randles circuit (a) and modifications including a CPE (b) or a Warburg element (c), considering Rs=100Ω, Rct=100Ω, β_{CPE}=0.85, A_W=0.01, C_{dl}=1µF. The arrows indicate an increase in frequency.

2.5.3 Circuits with two or three frequency relaxations

The following configurations are used to model data with two or three frequency relaxations, observed in the Nyquist plot as semicircles. Notice that the models with identical number of elements, produce the same exact frequency response, with the appropriate equivalent parameters for resistors and capacitors.

The circuit diagrams in Fig. 2.11 produce two frequency dispersions.

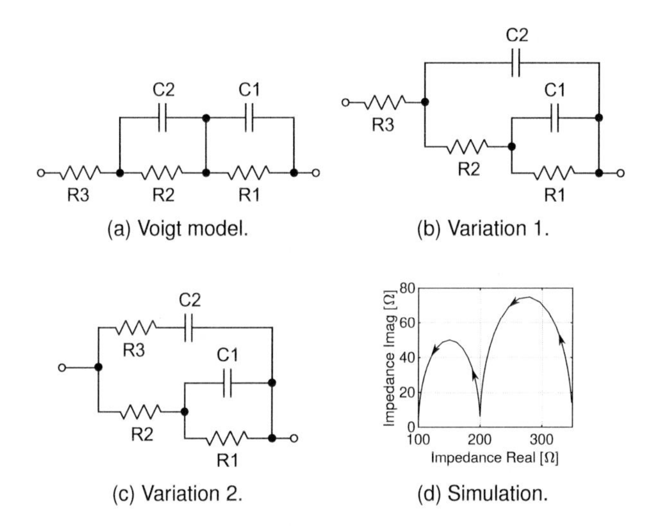

FIGURE 2.11: Circuit models with two dielectric relaxations.

The models in Fig. 2.12 exhibit three frequency relaxations.

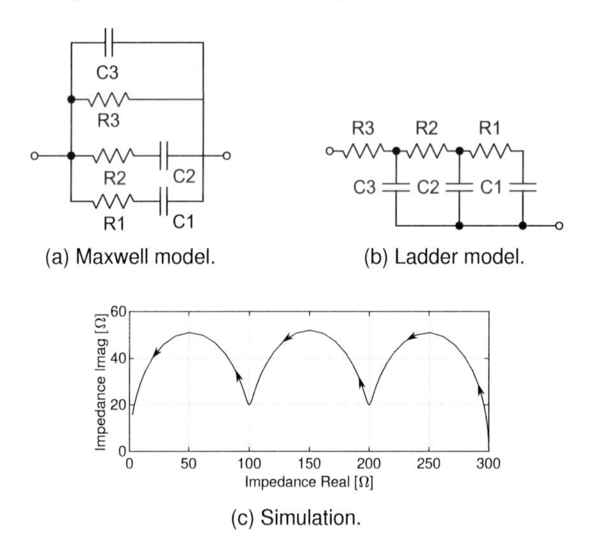

FIGURE 2.12: Circuit models with three dielectric relaxations.

2.5.4 Basic models for cell suspensions

The development of electrochemical models for single cells and cell cultures depends largely on the cell concentration and the experimental setup used for the measurement. The electrode configuration affects the impedance response, therefore a two-terminal setup requires an extra double-layer capacitance and some additional elements when compared to the four-terminal method.

One of the basic models to describe a cell suspension is proposed by Kyle [56] using a four-terminal setup. The model includes a resistor R_i for the cytoplasm, the resistance R_e of the extracellular medium and a capacitance C_m modeling the cellular membrane. The series resistance R_o is the impedance observed at high frequencies when the membrane capacitance becomes a short circuit. The model is observed in Fig. 2.13.

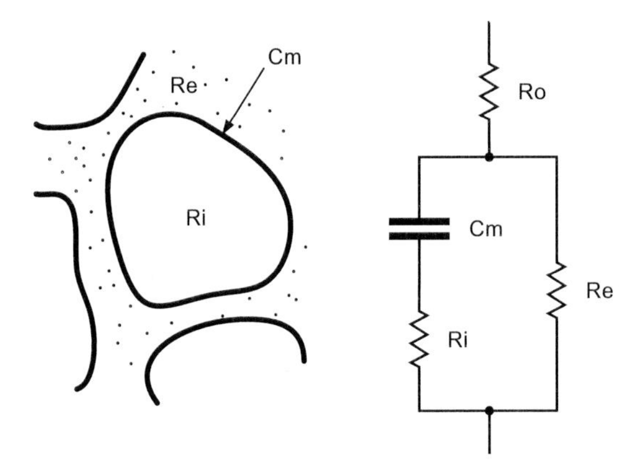

FIGURE 2.13: Cell model for a basic four-terminal cell experiment.

To illustrate the effect of the experimental setup, consider the single cell model of HeLa cells described by Wang [57]. The measurement setup consists on a two-electrode microfluidic chamber based on a coplanar waveguide. The model includes a single trapped cell and the solution.

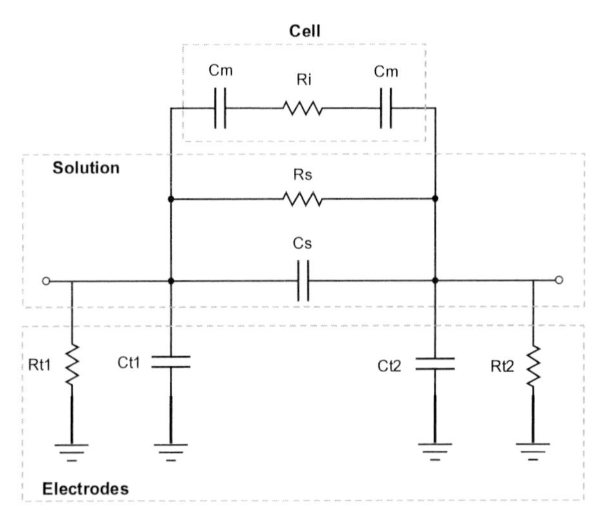

FIGURE 2.14: Cell model for a single HeLa cell trapped in a microfluidic device [57].

Other equivalent models study the effect of electrode shape on the impedance [58], ionic channels with a parallel R_{mem} component, and a separate membrane envelope for the cell nucleus. Additional studies may be consulted at [59–61].

2.6 Simulation of a 3D cell culture

The simulation of a cell culture may be approximated by a dielectric mixture: a set of spherical particles of permittivity ϵ_c suspended in a medium of permittivity ϵ_m.

The Maxwell-Garnett theory of effective medium approximations considers circular particles randomly located over an homogeneous extracellular medium [62]. The permittivity of the mixture is calculated as:

$$\epsilon_{mix} = \epsilon_m + 2p\epsilon_m \frac{\epsilon_c - \epsilon_m}{\epsilon_c + \epsilon_m - p(\epsilon_c - \epsilon_m)} \quad (2.15)$$

where ϵ_m is the effective complex permittivity of the extracellular medium, ϵ_c is the complex permittivity of the cells, and p is the area occupied by the cells.

If the volume fraction p is small, which means the cell concentration is low, the Maxwell-Wagner equation is used [63]:

$$\epsilon_{mix} = \epsilon_m \frac{(2\epsilon_m + \epsilon_c) - 2p(\epsilon_m + \epsilon_c)}{(2\epsilon_m + \epsilon_c) + p(\epsilon_m + \epsilon_c)} \quad (2.16)$$

And for higher volume fractions where the sample exhibits a higher concentration, the Maxwell-Wagner-Hanai formula [64][65] is used:

$$\frac{\epsilon_{mix} - \epsilon_c}{\epsilon_m - \epsilon_c} \left(\frac{\epsilon_m}{\epsilon_{mix}} \right)^{1/3} = 1 - p \quad (2.17)$$

The permittivity of the cells ϵ_c in the above equations may be computed by using a simple sphere, a single-shell model or a double-shell model. These approximations give different levels of detail to the cells. In addition, insertions in some of the layers may be used to describe additional organelles. This is shown in Fig. 2.15.

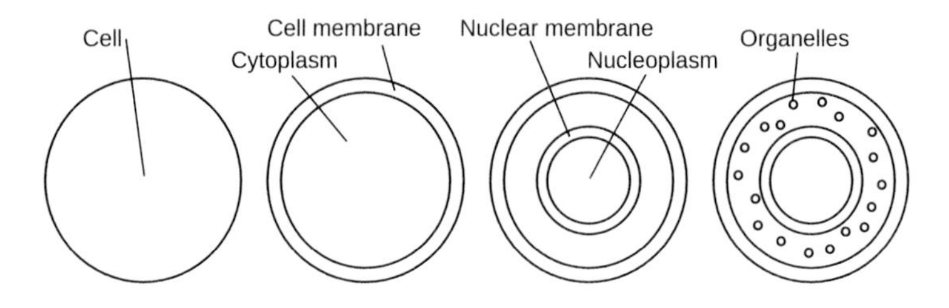

FIGURE 2.15: Simple sphere, single-shell, double-shell and organelle models [54].

A mathematical model was developed by Edgar Salazar in COMSOL Multiphysics® using the single-shell approach [66]. The objective of this simulation is to calculate the total effective permittivity between the parallel plates, as a function of frequency.

Geometry: The model automatically generates a suspension of cells, randomly located in a measurement chamber with parallel-plate electrodes. The input parameters for the geometry generation are the cell radius and number of cells. A model with 100 cells is shown in Fig. 2.16.

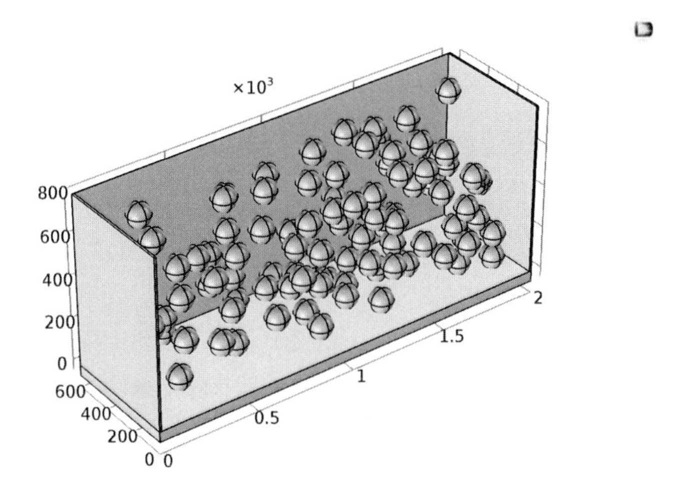

FIGURE 2.16: Geometry of the COMSOL® cell culture model with 100 cells.

Parameters: The permittivity and conductivity of the model elements are set according to the values found in Table 2.3, which includes common physical constants for lymphocytes [67].

TABLE 2.3: Parameters of the materials used in the model [67].

Element	Permittivity	Conductivity
Nucleoplasm	120	0.95
Nuclear envelope	41	0.003
Cytoplasm	60	0.48
Membrane	5.8	8.7e-6
Extracellular medium	65	0.16

The beta dispersion is modeled by using Eq. 2.16, and the gamma dispersion is modeled by means of the Debye relaxation from Eq. 2.7.

The simulation results are presented in Fig. 2.17. The permittivity and conductivity change in steps, showing the beta and gamma frequency dispersion regions. This figure shows that the impedance of a cell culture changes in steps, and that this behavior is explained due to the mixture of dielectric properties.

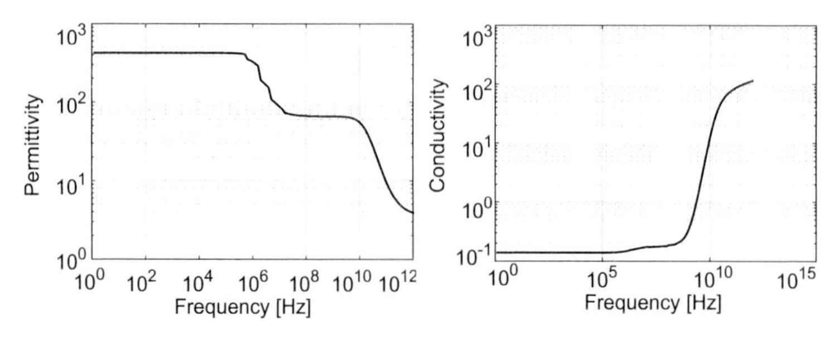

FIGURE 2.17: Results of the simulation of a cell culture in COMSOL®.

In a physical experiment with actual cells, it is expected to obtain more dispersions, since cells have different membranes and organelles that would produce an even more detailed impedance pattern.

Experimental impedance curves are fitted to equivalent circuit models by using numerical approximations. The following sections describe a fitting toolbox which was developed as part of this work, based on algorithms freely available on MathWorks® MATLAB® Central™.

2.7 Commercial tools for fitting

The circuit models from the previous section are stored as a netlist, which is a textual description of the circuit elements and their interconnections. Circuit simulators take the netlist and ask the user to assign numerical values to each component in the circuit, to produce the frequency response output. This is done by calculating the transfer function $H(\omega)$ which describes the impedance of the system for every frequency.

Fitting is exactly the opposite. The measured impedance response of the physical system is available, and the optimizer tries to guess the circuit element values by iterative methods. On the first iteration, the initial values are used to compute the simulated frequency response, and an error function compares this to the measured frequency response. The optimizer modifies the parameters, computes another simulation and compares the error with the previous iteration. The process is repeated until the error is acceptable or the maximum number of iterations is reached. The processes of simulation and fitting are shown in Fig. 2.18.

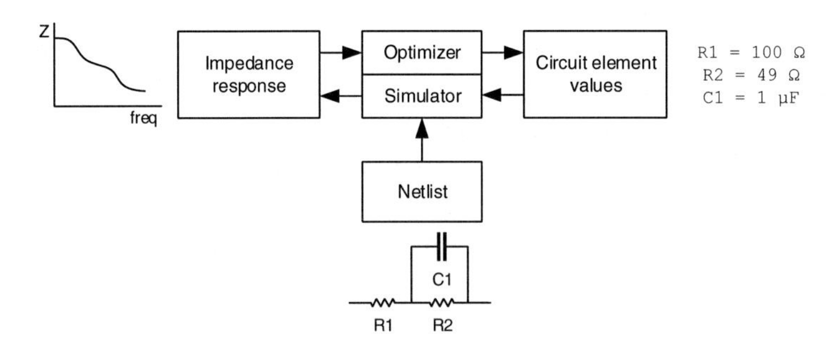

FIGURE 2.18: Flow diagram of the fitting and simulation processes.

Many software tools for impedance fitting already exist in the market. Programs are very different in terms of the fitting algorithms, weighting functions, statistical analysis and error estimates calculation. A list of software for fitting is compiled in Table 2.4.

TABLE 2.4: Software tools for fitting EIS data to equivalent circuit models.

Company	Software name	Version	Release	Ref
J.L. Dellis	ZFit	1.2	19.05.2010	[68]
J.R. Macdonald	LEVMW	8.13	05.2015	[69][70]
S. Koch	Elchemea Analytical	1.5.1	28.09.2017	[71]
A. Bondarenko	EIS Spectrum Analyzer	1.0	16.07.2013	[72][73]
Gamry Inc.	Echem Analyst	6.33		[74]
Gamry Inc.	Echem Analyst	7.05		[74]
Metrohm Autolab	NOVA	2.1.2	05.05.2017	[75]
BioLogic	EC-Lab	11.16	12.08.2017	[76]
RHD Instruments	RelaxIS	3.0.5		[77]
Kumho Petrochemical	MEISP	3.0	03.2002	[78]
EChem software	ZSimpWin	3.10d	15.02.2004	[79]
B. Boukamp	EQIVCT			[80][81]
Novocontrol	WinFIT			[82]
Zivelab	Zman	2.4	13.10.2014	[83]
Ivium Technologies	IviumSoft			[84]
Scribner Associates	ZView	3.5d	13.09.2017	[85]

A detailed implementation of a fitting toolbox for MATLAB® based on open-source scripts and functions is described in the following section.

2.8 Impedance fitting toolbox

A MATLAB® toolbox was developed for fitting impedance spectroscopy data to equivalent circuit models. The toolbox is presented in Fig. 2.19, showing the main dialog on the left side, and the Nyquist plots of the measured and fitted impedance on the right side. The user interface is shown in Fig. 2.19 featuring the Bode plot of experimental data and the predefined equivalent circuit models. The main window labeled 'eistoolbox' includes four sections: input data management, circuit model parameters, fitting configuration and output data management.

The input data section allows loading experimental data files formatted as CSV or Gamry DTA. The user can plot the files as Bode or Nyquist. The circuit parameters section includes the equivalent circuit represented as a fitting string, initial values and boundary conditions. The fitting configuration allows selecting the algorithm and weighting, number of iterations and number of cores for parallel fitting. The output data section allows the user to export the fitted parameters.

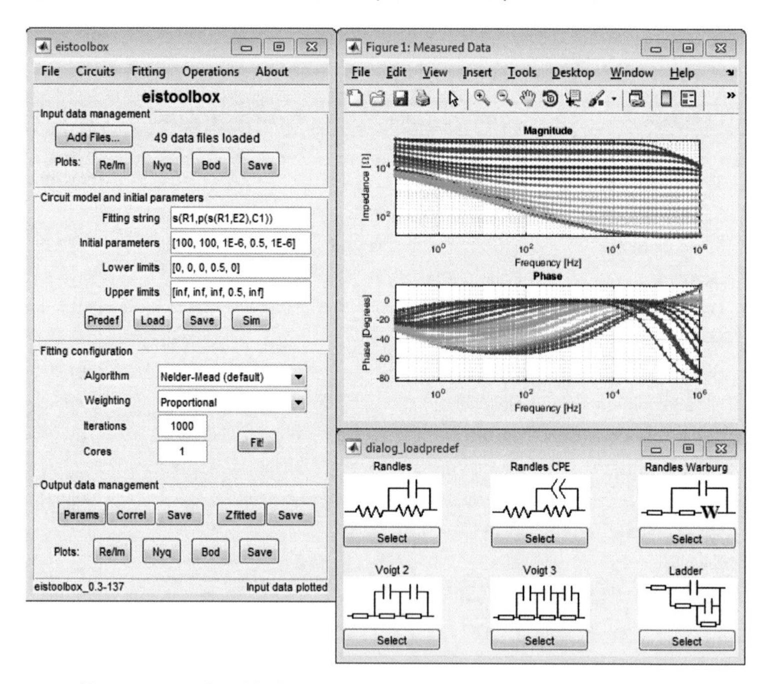

FIGURE 2.19: Graphical user interface of the fitting toolbox for MATLAB®.

2.8.1 Fitting algorithms

The fitting parameters can be found using non-linear constrained optimization algorithms. The most common for EIS are the Levenberg-Marquardt [86] and the Nelder-Mead [87] algorithms, as they are widely used in commercial software tools, although there are many others. For example, a comparison of 21 different optimizer functions for MATLAB® is found in the iFit library [88], developed under the EUPL license. The optimizers vary in terms of the required number of iterations and the performance under noisy conditions.

Currently the toolbox uses the fminsearchbnd algorithm by John D'Errico [89], which is based on the Nelder-Mead method with boundary conditions for each individual parameter. The fitting engine of the toolbox is an adaptation from the Zfit script by Jean-Luc Dellis [68], available for download from the MathWorks® MATLAB® Central™ repository.

The objective of numerical optimization is to minimize the distance function, defined as the sum of squares between the experimental and the fitted data. The original distance function was removed from Zfit and implemented as described by A. Lasia [90] for improved control on the weighting types, as shown in the following equation:

$$\mathrm{dist} = \sum_{i=1}^{n} w_i'(Z_{i,exp}' - Z_{i,fit}')^2 + w_i''(Z_{i,exp}'' - Z_{i,fit}'')^2 \tag{2.18}$$

where Z_{exp} is the measured impedance, and Z_{fit} is the simulated impedance with the fitting parameters of the last iteration.

The parameters w_i' and w_i'' are the weighting values for the real and imaginary parts, and can be selected according to the values in Table 2.5.

TABLE 2.5: Weighting types for function minimization.

Weighting	**Equation**
Unit	$w_i' = w_i'' = 1$
Modulus	$w_i' = w_i'' = \|Z\|$
Proportional	$w_i' = Z_i', w_i'' = Z_i''$

In addition, multicore support has been added by means of the Parallel Computing Toolbox™ which supports up to 12 cores for simultaneous fitting. This is particularly useful when a large set of experiments needs to be processed in batch.

2.8.2 Circuit strings and predefined models

Custom circuits are entered in a text box, formatted as a string of parallel or series combinations of circuit elements using the s() and p() operators, with any number of free parameters. The available circuit elements and labels are given in Table 2.6. The label includes a letter for parameter identification, and a number representing how many free parameters the element expects. These free parameters are called P and Q in the table below, and are the outputs of the model.

TABLE 2.6: Equivalent circuit elements and their MATLAB® implementation.

Label	Element	Equation
R1	Resistor	Z(f) = P
C1	Capacitor	$Z(f) = 1/j2\pi fP$
L1	Inductor	$Z(f) = j2\pi fP$
E2	Constant Phase Element	$Z(f) = 1/[P(j2\pi f)^Q]$

For example, for a Randles model with Warburg (see Fig. 2.10c), the circuit string is written as s(R1,p(s(R1,E2),C1)). The elements are the series resistance (Rs), charge transfer resistance (Rct), double-layer capacitance (Cdl) and the Warburg element (W), which is obtained by setting Q=0.5 in the CPE element.

In addition, the toolbox includes predefined circuit models, presented in Fig. 2.20. These can be loaded by pressing the 'Predef' button and may help the user to understand the syntax and the use of initial values, lower and upper boundary limits. The tool also supports saving and loading additional circuit models as text files with the .ckt file extension.

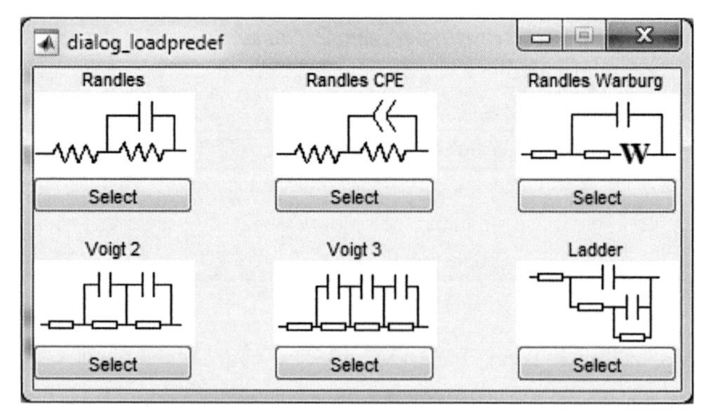

FIGURE 2.20: Equivalent circuit models implemented in the toolbox.

2.8.3 Error estimation

The tool calculates coefficients of determination and chi-square goodness-of-fit, using the experimental data and the simulated impedance of the fitted circuit. This is a significant addition to the original Zfit script, which included no option for validation of the fitting results. The error estimation information is presented in Fig. 2.21.

The plots in the upper part of this figure contain the fitted impedance in the vertical axis, and the measured impedance in the horizontal axis. A perfect match of data points results in a straight line. The lower part of the figure includes the numeric coefficients of determination and the chi-square parameter.

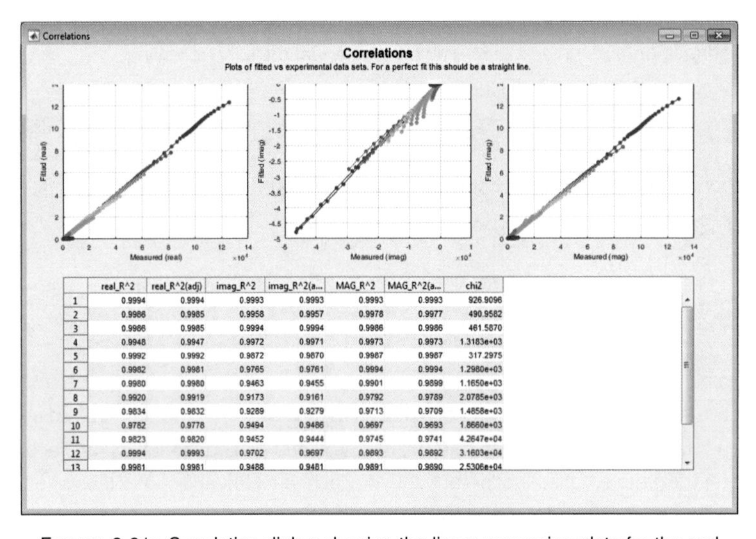

	real_R^2	real_R^2(adj)	imag_R^2	imag_R^2(a...	MAG_R^2	MAG_R^2(a...	chi2
1	0.9994	0.9994	0.9993	0.9993	0.9993	0.9993	926.9096
2	0.9986	0.9985	0.9958	0.9957	0.9978	0.9977	490.9582
3	0.9986	0.9985	0.9994	0.9994	0.9986	0.9986	461.5870
4	0.9948	0.9947	0.9972	0.9971	0.9973	0.9973	1.3183e+03
5	0.9992	0.9992	0.9872	0.9870	0.9987	0.9987	317.2975
6	0.9982	0.9981	0.9765	0.9761	0.9994	0.9994	1.2980e+03
7	0.9980	0.9980	0.9463	0.9455	0.9901	0.9899	1.1650e+03
8	0.9920	0.9919	0.9173	0.9161	0.9792	0.9789	2.0785e+03
9	0.9834	0.9832	0.9289	0.9279	0.9713	0.9709	1.4858e+03
10	0.9782	0.9778	0.9494	0.9486	0.9697	0.9693	1.8660e+03
11	0.9823	0.9820	0.9452	0.9444	0.9745	0.9741	4.2647e+04
12	0.9994	0.9993	0.9702	0.9697	0.9893	0.9892	3.1603e+04
13	0.9981	0.9981	0.9488	0.9481	0.9891	0.9890	2.5306e+04

FIGURE 2.21: Correlation dialog showing the linear regression plots for the real part, imaginary part and magnitude of the impedance.

The coefficient of determination, R^2, describes how closely the fitted results are matching the original impedance data. This is done independently for the magnitude, real and imaginary parts of the impedance. These coefficients are computed as described in the MathWorks® website [91] by means of the following equations:

$$R^2 = 1 - \frac{SS_{resid}}{SS_{total}} \tag{2.19}$$

$$SS_{resid} = \sum(\text{observed} - \text{fitted})^2 \tag{2.20}$$

$$SS_{total} = (n-1) \times \text{Var}(\text{observed}) \tag{2.21}$$

The coefficient of determination is adjusted to include the degree of the regression:

$$R^2_{adj} = 1 - \frac{SSresid}{SStotal} \cdot \frac{n-1}{n-d-1} \tag{2.22}$$

where d is 1 for linear regression, 2 for quadratic regression, 3 for cubic regression. The tool uses linear regression coefficients only.

Another measurement of the error of the fitting is obtained by the chi-square goodness-of-fit parameter (χ^2). The general equation is the following:

$$\chi^2 = \sum \frac{(o-e)^2}{e} \tag{2.23}$$

Where *o* represents the observed (fitted/simulated) data and *e* represents the expected (measured) data.

2.8.4 Fitting example

A cubic chamber with volume of 1 cm^3 was designed with stainless steel electrodes at two sides. The surface area of each electrode is of 1 cm^2. An aqueous solution of 0.6 M $CaCl_2$ was prepared in the laboratory, and diluted by half with deionized water, down to a concentration of 0.2 mM. Each sample is placed in the chamber and the impedance of each sample is measured from 0.1 Hz to 1 MHz using a Gamry 1000 Interface, using a two-terminal experiment configuration.

The data is modeled using the Randles model with a Warburg element, shown in Fig. 2.22, since the electrolyte contains dissociated ions. The Warburg element describes the diffusion of these ions in the bulk electrolyte and therefore it is the optimal choice for this experiment.

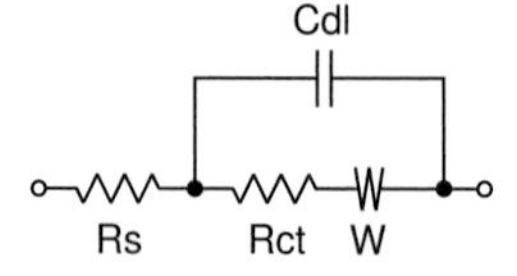

FIGURE 2.22: Randles model with Warburg element.

After fitting the data, the fitting parameters are shown in Table 2.7.

TABLE 2.7: Fitting results for $CaCl_2$ samples using the Randles Warburg model.

Concentration (mmol/L)	Rs (Ω)	Rct (Ω)	W (Ω)	Cdl (F)
0.03	1.29E+04	2.19E+04	3.16E-05	5.97E-11
0.07	1.29E+04	7.56E+03	4.77E-05	3.07E-10
0.15	1.14E+03	1.90E+04	4.81E-05	1.30E-11
0.29	4.16E-10	1.03E+04	5.73E-05	1.49E-11
0.59	6.58E+03	1.19E+02	5.55E-05	5.11E-07
1.17	1.56E-09	3.17E+03	7.06E-05	1.07E-11
2.34	1.58E+03	5.11E+03	1.04E-04	7.35E-06
4.69	6.51E+02	1.73E+02	1.71E-04	2.59E-10
9.38	5.32E+02	2.22E+04	1.55E-04	2.75E-04
18.75	2.52E+02	1.64E+03	2.75E-04	2.10E-05
37.50	1.17E+02	1.11E+02	2.75E-04	3.87E-06
75.00	8.10E+01	5.60E+04	2.29E-04	3.71E-04
150.00	4.51E+01	1.15E+03	2.15E-04	2.17E-05
300.00	2.50E+01	1.40E+04	7.11E-05	2.69E-04
600.00	1.91E+01	3.37E+03	8.85E-05	1.30E-04

The value of the solution resistance (Rs) is above 10 kΩ for the lowest concentrated sample, and it decreases as the concentration of ions in the sample increases, as shown in Fig. 2.23a. This was the expected behavior. The double-layer capacitance (Cdl) in Fig. 2.23b shows the opposite: for low ionic concentration this capacitance is small, and it increases as ions accumulate at the interface.

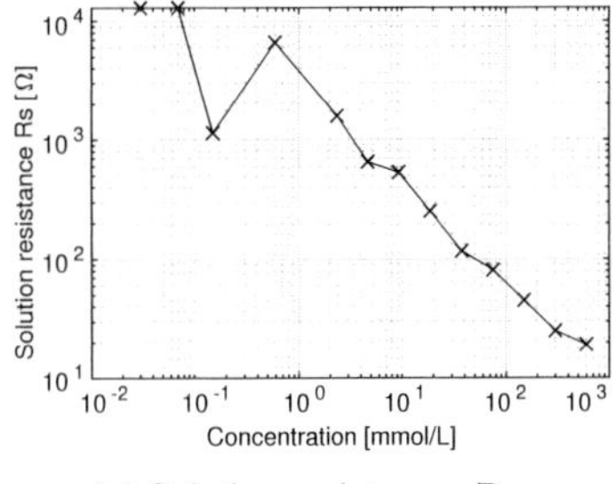

(a) Solution resistance, Rs.

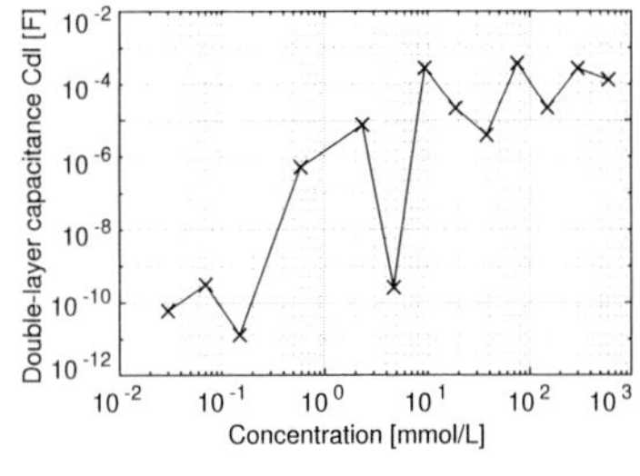

(b) Double-layer capacitance, Cdl.

FIGURE 2.23: Model parameters plotted as a function of $CaCl_2$ concentration.

The error information is given in Table 2.8, where most of the samples have a high coefficient of determination (above 0.9) meaning that the difference between experimental and fitted values is small.

TABLE 2.8: Coefficients of determination and chi-square errors.

Concentration (mmol/L)	R^2 real	R^2 imag	R^2 mag	χ^2 mag
0.03	0.9121	0.6859	0.9490	334440.01
0.07	0.7471	0.3300	0.8329	584500.0
0.15	0.9948	0.9972	0.9973	1318.7
0.29	0.9992	0.9872	0.9987	317.3
0.59	0.9955	0.9472	0.9986	6149.4
1.17	0.9980	0.9463	0.9901	1165.0
2.34	0.9890	0.9777	0.9935	6569.1
4.69	0.9835	0.9289	0.9714	1480.1
9.38	0.5639	0.7471	0.7133	19027.0
18.75	0.9845	0.9875	0.9910	2883.4
37.50	0.9997	0.9712	0.9962	223.9
75.00	0.6243	0.7222	0.8136	15903.0
150.00	0.9969	0.9715	0.9962	1885.8
300.00	0.7090	0.6686	0.8347	19472.0
600.00	0.8013	0.6594	0.8583	27133.0

2.9 Chapter conclusion

Impedance spectroscopy is measurement technique that measures the permittivity and conductivity of the samples over a specific range of frequencies. Experiments may be performed using two, three or four terminals to improve measurement accuracy and reduce parasitic effects such as the double-layer effect.

Modeling of electrochemical systems is achieved by using equivalent circuit elements such as resistors, capacitors and inductors. The experimental impedance data is processed by an iterative fitting algorithm to find the individual values of the components. A MATLAB® tool was developed and used for fitting experimental data with a high level of accuracy, as shown by the regression coefficients from Table 2.8. This tool allows fitting experimental data to equivalent circuit models with any number of parameters, and it is used to fit experiments in the next chapter.

Chapter 3

Applications of impedance spectroscopy

This chapter covers selected applications of impedance spectroscopy. Section 3.1 includes a test experiment to differentiate cancerous and healthy tissues in mice. Section 3.2 describes experiments using samples of liquids infected with viruses. Section 3.3 describes the constant monitoring of fruit during ripening. Section 3.4 includes experiments to measure cell viability, the concentration of alive cells in a given cell culture. Section 3.5 includes a study aimed to the measurement of blood lactate concentration in athletes by using an impedance sensor.

3.1 Impedance response of tumor and healthy cells

This section describes impedance spectroscopy experiments performed on mice, two weeks after being implanted with human tumor xenografts. The measurements were carried out at the Institute of Anatomy and Experimental Morphology of the University Medical Center Hamburg-Eppendorf (UKE) using a Gamry Interface 1000 potentiostat in the frequency range from 1 Hz to 1 MHz. The goal is to observe the impedance of cancerous and non-cancerous tissues in mice, and confirm if cancer can be detected from impedance measurements.

Background: A large number of diseases are developed at cellular levels, and the most common is cancer. It originates when cells mutate and start dividing in an uncontrolled way, forming an agglomeration known as a tumor. Benign tumors are localized at a specific part of the body, whereas malignant tumors can metastasize and spread to other parts and invade healthy organs. There are more than one hundred types of cancer [92], which are classified depending on the region where it is originated (topography) and the cell types affected (morphology) [93]. Based on the site of origin, cancers can be classified as carcinoma (from epithelial cells), blastoma (from embryonic cells), leukemia and lymphoma (from blood cells) or sarcoma (from connective and supportive tissue).

Detection of cancer at early stages can save lives, since some malignant tumors may be surgically removed before metastasis occurs. The traditional methods for early cancer detection are computed tomography (CT) and magnetic resonance imaging (MRI). Nuclear imaging tools are positron emission tomography (PET) and single-photon emission computer tomography (SPECT). Periodical screening by CT and MRI is recommended only for persons with genetic risk factors, since the radiation may have negative health effects. These tools can detect tumors with a minimum size of millimeters, known as the detection limit [94][95].

Electrical screening techniques such as electrical impedance tomography (EIT) have been recently developed as an alternative to conventional tomography. This technique measures the dielectric properties of tissues and it is based on electrochemical impedance spectroscopy (EIS) and electric cell-substrate impedance spectroscopy (ECIS).

Materials and methods: For this study we use two immunodeficient mice from the RAG2 line (Mice 1-2, with tumor) and one mouse of the line PFP/RAG2 (Mouse 3, healthy). Mice are genetically altered to develop immunodeficiency by blocking the Recombination Activation Gene 2 (RAG2) and PFP genes. The RAG2 gene inhibits the maturation of T and B lymphocytes [96], and the PFP gene produces a depletion of the NK cells [97]. In combination, the immune system is unable to respond to tumor formation, and the implanted tumor may spread to other organs.

The tumor cells are xenographs from human squamous cell carcinoma, and the specific cell lines are UT-SCC-5 (procedent from head and neck) and UM-SCC-10A (from laryngeal carcinoma). The tumors may be extracted from a human patient and then implanted in the mice as described in [98], or as in this case, using commercially available cell cultures.

A summary of the mice and tumor cell lines is given in Table 3.1.

TABLE 3.1: Summary of mice and tumor cell lines used in the experiment.

Mice	Mice Line	Tumor Cell Line
Mouse 1	RAG2	UT-SCC-5
Mouse 2	RAG2	UM-SCC-10A
Mouse 3	PFP/RAG2	Healthy

The tumor is implanted and allowed to grow until it reaches 10% of the total mouse weight. This typically occurs between two and three weeks after tumor implantation. When the tumor reaches a target size, the mouse is killed and biological samples are extracted. The scientists at the UKE extract blood and bone marrow. Afterwards, the organs are dissected and placed into small containers where electrodes are placed with a separation distance of approximately 2 mm. The electrodes have a needle which is inserted directly into the tissue. Electrodes are then placed into tissues in the brain, lungs, heart, liver, leg muscle and the tumor itself.

Experimental results: The impedance of each tissue sample is recorded using the Gamry Interface 1000 in potentiostatic mode, applying an AC voltage of 100 mV, sweeping the frequency from 1 Hz to 1 MHz, using 10 points per decade. The magnitude of impedance is shown in Fig. 3.1, and the phase in Fig. 3.2.

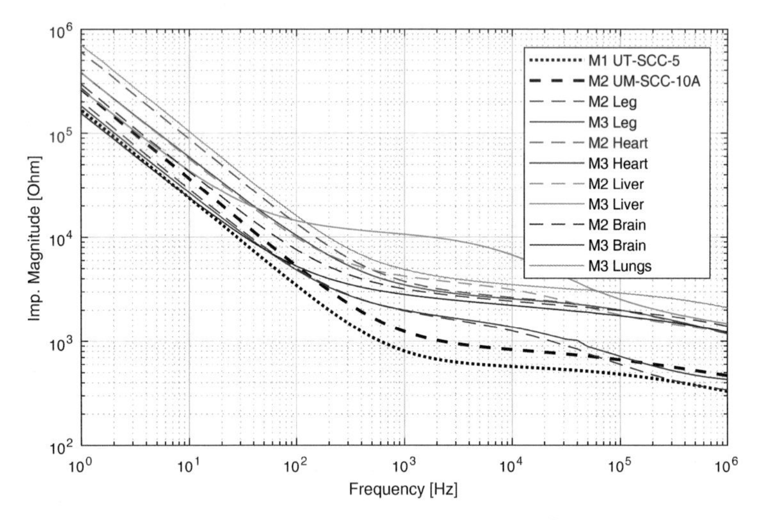

FIGURE 3.1: Impedance magnitude of different mice tissues. Dotted lines are from the first mouse (Mouse 1) with the UT-SCC-5 tumor, dashed lines are from mice with an implanted UM-SCC-10A tumor (Mouse 2), and continuous solid lines are from a healthy mouse with no tumors implanted (Mouse 3).

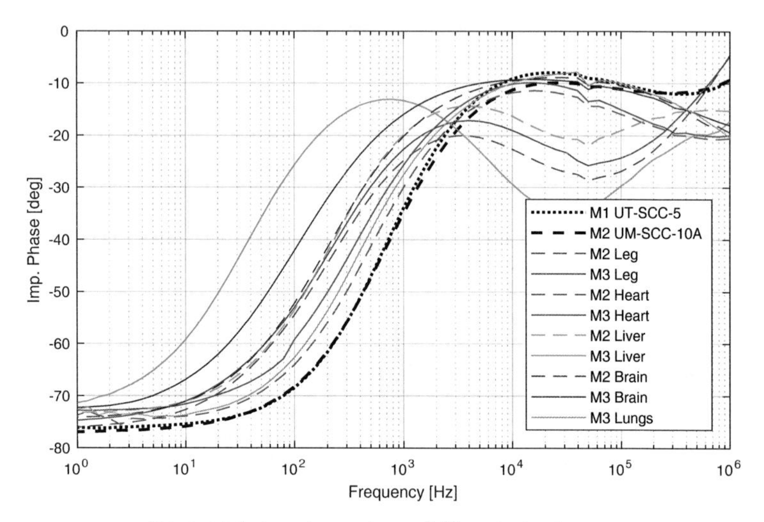

FIGURE 3.2: Impedance phase of different mice tissues.

It was found that the impedance of tumor tissue is lower than the impedance of other healthy tissues. The hypothesis from literature is that cancer tissues contain a high number of non-viable cells which already have a weakened cell membrane, or it has already failed to protect the cell, reducing the membrane capacitance considerably. This effect has been observed on liver tissues [24], prostate [99][100], breast [101] and skin [102].

The impedance difference within the same type of tissue is most evident in the liver, and it can be appreciated as well in the leg muscles, where the impedance for a cancerous mouse (dashed green) is lower than the reference measurement (solid green) for a healthy mouse. This suggests cancer has already spread to the liver and the legs.

The phase shift observed in Fig. 3.2 occurs also in higher frequencies as compared with the rest of the studied biological material. This is the case for both cell lines of human cell carcinoma.

Results show a strong alpha dispersion, reflected on the linear increase of the impedance magnitude for lower frequencies. The phase diagram shows a strong capacitive effect with a phase angle of almost -80° for frequencies close to DC. These effects are very similar to the double-layer effect observed in aqueous electrolytes.

The alpha dispersion effect in muscular tissue is explained by Schwan [48] as follows: "in muscle tissue the existence of the sarcoplasmic reticulum appears to be primarily responsible for the strong dispersion". The sarcoplasmic reticulum is a muscle cell organelle that replaces the endoplasmic reticulum of other types of cells. This structure has the main function of storing calcium ions. Therefore, these ions also form a double-layer effect. This parasitic effect could be removed if a four-terminal electrode configuration is used.

In addition to the Bode response, the Nyquist diagram is also plotted. Due to the large variation of impedances, both axes are plotted in logarithmic scale. The Nyquist diagram shows a clear difference between each of the studied tissue types.

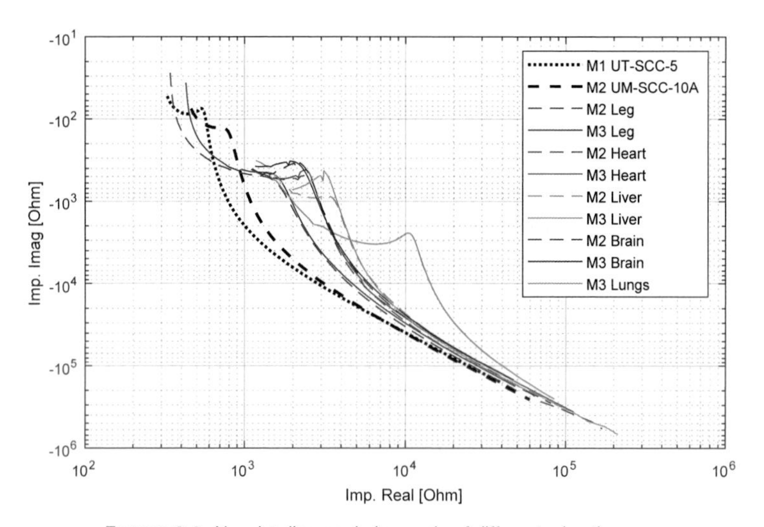

FIGURE 3.3: Nyquist diagram in log scale of different mice tissues.

Conclusion: The study of different tissues from the mice showed that it is possible to identify cancer cells, as these type of biological material showed the lowest impedance when compared to the rest of the samples. This is indeed the working principle of current electric impedance tomography devices (EIT) which plot a two-dimensional surface plot of the local changes of the impedance.

This measurement also demonstrated that each individual tissue exhibits a specific impedance pattern, and the impedance response for each tissue is almost the same in both the RAG2 (Mouse 2) and PFP/RAG2 (Mouse 3) immunodeficient mice, with the exception of the liver tissue where the difference is more noticeable.

3.2 Liquid samples infected with bacteria

In this section we analyze samples of a cell culture originated from the eye of patients with a bacteria infection. The basis of this theory is that every cell has a specific impedance 'fingerprint' which can be used to differentiate one cell type from another, and in this case, the samples containing bacteria would have a distinct impedance pattern when compared to healthy samples.

Materials and methods: The samples used for this section were provided by the University Medical Center Hamburg-Eppendorf (UKE). The volume of these kind of samples is very limited, since the liquid extracted from the eye is only one or two drops. Only one drop was already enough to perform the measurements.

The measurement electrodes consisted of a plastic membrane with a thin coating of electrodeposited gold, forming a pair of interdigitated electrodes. These electrodes were produced in the company i3membrane GmbH. The exact geometry and specifications of these electrodes cannot be disclosed, since the design is protected by a non-disclosure agreement. A drop of the liquid sample is placed on top of the electrode, and the impedance is measured immediately after the drop is deposited. The electrode is discarded afterwards.

Experimental results: Impedance was recorded with the Agilent 4284A LCR meter. A total of six samples was analyzed. Two samples from infected patients, two samples from healthy persons, and two blood samples for comparison. The averaged results are presented in Fig. 3.4 and Fig. 3.5.

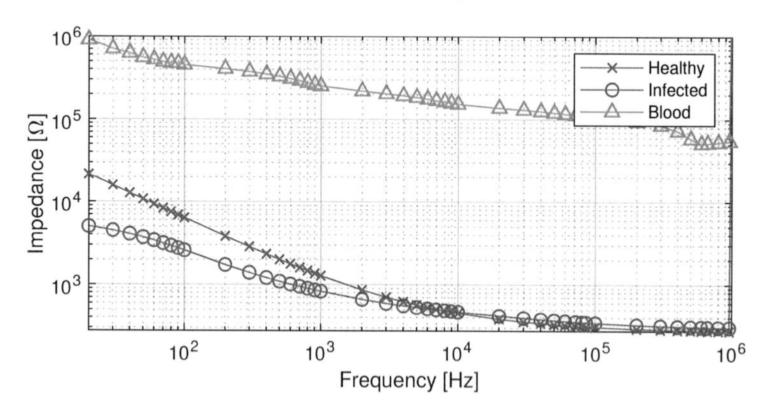

FIGURE 3.4: Impedance magnitude of healthy and infected samples, compared with the impedance of blood.

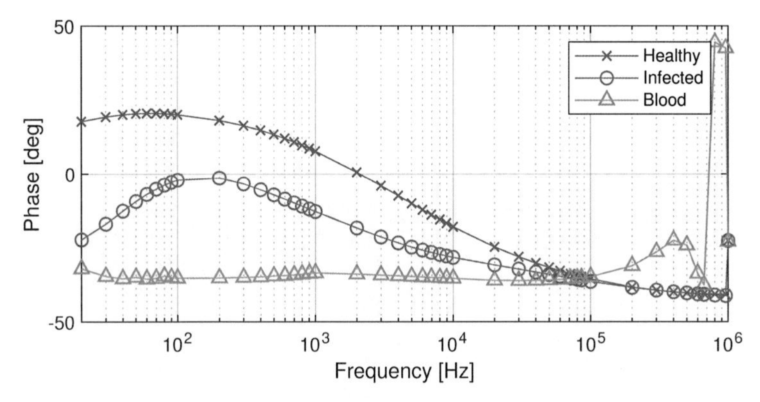

FIGURE 3.5: Impedance phase of healthy and infected samples, compared with the phase of blood.

Conclusion: This experiment demonstrated that it is possible to identify the infected samples by observing the impedance spectrum, since results showed a clear difference between healthy and infected samples. An infected sample contains bacteria and other substances that increase the conductivity. This is confirmed also by the measurement of blood samples, which exhibited the largest impedance on the plots since it contains a large number of cells and particles that limit the current flow. If a large collection of data is obtained and statistically processed, this could be used to build an analytical system capable of detecting the presence or absence of bacteria infections, without the need of any specific biological test.

3.3 Fruit ripening

The parameters for assessing fruit maturity at harvesting time are firmness, color, size and shape, which are subjective and performed by manual inspection. In this section, impedance spectroscopy is used to monitor the maturity and ripening process of fruits, to detect the quality of the food, or estimate how many days have passed since the harvest.

The process of ripening in food modifies electrical properties such as conductivity and permittivity. This is partially explained by the loss of water during ripening, and also due to the existence of electrochemical reactions that modify the biochemical structure of the food itself. This method has already been explored with banana [15], mango [16, 17] and kiwi [18] among other variety of fruits and vegetables.

Materials and methods: A fresh strawberry from the supermarket was placed in a special electrode setup, as shown in Fig. 3.6. The electrodes are four screws which are arranged in such a way that enable holding the fruit over several days. These screws are contacted in pairs by two wires, which are connected to the Agilent 4285A LCR Meter, in the frequency range from 75 kHz to 30 MHz.

(a) Day 1. (b) Day 4. (c) Day 10.

FIGURE 3.6: A strawberry placed on top of four electrodes for impedance sensing.

The LCR meter is connected to a computer running MATLAB®, and a script was programmed to sample one impedance point every 30 minutes, repeating the measurements until the script is manually stopped. The system was left running in the laboratory with a glass cover. The impedance is monitored for ten days.

Experimental results: The impedance results are shown in Fig. 3.7 for both magnitude and phase. The first recorded curve is shown in blue, and each measurement is indicated by a color progression towards red.

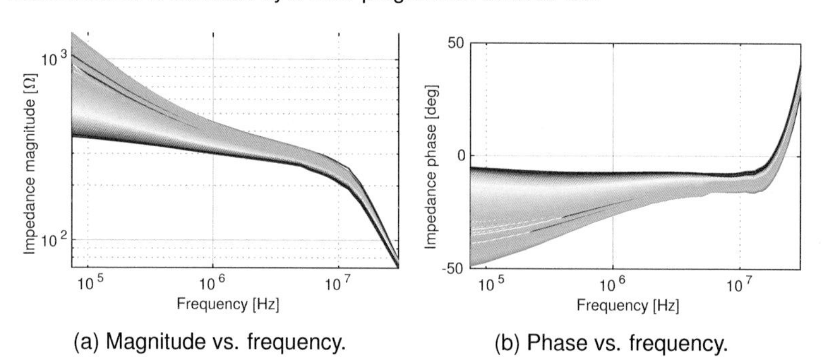

(a) Magnitude vs. frequency. (b) Phase vs. frequency.

FIGURE 3.7: Impedance of the strawberry as a function of frequency over ten days. The blue lines are recorded on day 1, and the red lines are from day 10.

The impedance curves from Fig. 3.7 are evaluated at frequencies of 100 kHz, 500 kHz, 1 MHz and 10 MHz. The impedance is plotted against the time in Fig. 3.8. This shows that the magnitude of the fruit increases constantly until a peak is reached, approximately after two days (48 hours) and then it shows a strong decline. The same behavior was observed in mangoes by M. Rehman [16]. The phase starts decaying to an almost pure resistive behavior after ten days.

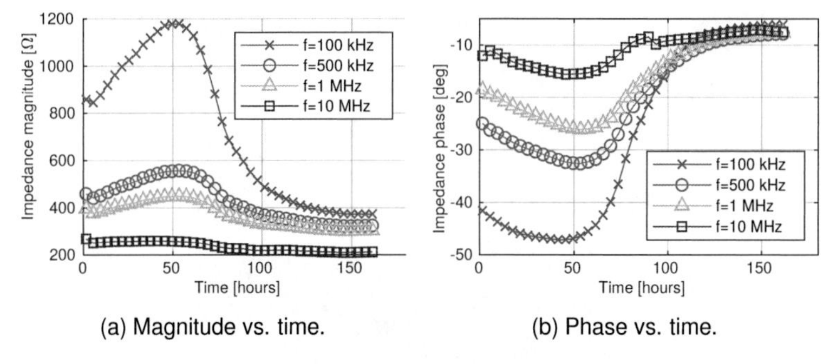

(a) Magnitude vs. time.

(b) Phase vs. time.

FIGURE 3.8: Impedance of the strawberry as a function of time over ten days, recorded at 100 kHz.

Conclusion: It was found that the impedance of the strawberry increases during the beginning of the ripening process, as in Fig. 3.8a, and it decreases later until a minimum value is reached. This is consistent with the experimental results consulted from literature, especially in agreement with [16]. The hypothesis is that chemical reactions and loss of water occur during ripening, modifying the conduction pathways in the fruit tissue. The method could be further used to study the effectiveness of different preservatives used in the food industry.

3.4 Cell viability

The viability of a cell is defined as the percentage of living cells present in a cell culture, as contrasted by the cells that are damaged or die during the experiment as consequence of contamination, field exposure or the change in the environmental properties such as temperature, pH or carbon dioxide concentration. In this section we observe the variations of the impedance of a cell culture over time, by placing the cells on a special chamber without any method to preserve cell vitality. After four hours, most of the cells would be already dead and the impedance could give evidence of this condition.

Materials and methods: A culture of CHO DP-12 cells was obtained from the Institute of Bioprocesses and Biosystems Engineering of the TUHH. The cells are placed in the experimental chamber of Fig. 3.9. The electrodes are connected to the Gamry Interface 1000E potentiostat, and the impedance is monitored by using a stimulation voltage of 50 mV over the frequency range from 0.1 Hz up to 1 MHz. An impedance curve is recorded every five minutes, over the course of four hours.

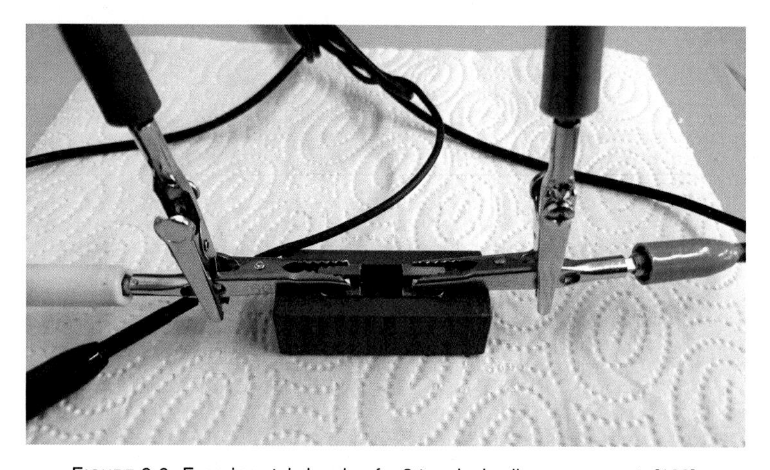

FIGURE 3.9: Experimental chamber for 2-terminal cell measurements [103].

Experimental results: The impedance of the cell culture is plotted at the beginning of the experiment, after ten minutes, and after four hours. These results are shown in Fig. 3.10. The absence of temperature and atmosphere controls result in cell death after four hours. It is observed that the impedance of cells decreases as time progresses, resulting in a 50% impedance reduction after cell death.

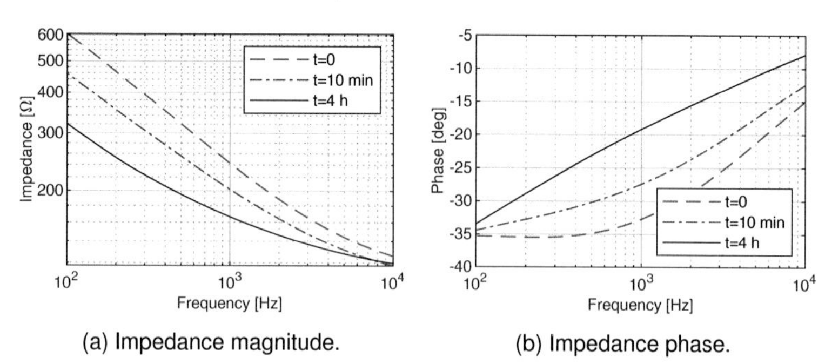

(a) Impedance magnitude. (b) Impedance phase.

FIGURE 3.10: Impedance of a CHO DP-12 cell culture, from 100 Hz up to 10 kHz.

The phase measured over the complete frequency range of the device is plotted in Fig. 3.11. The frequencies of interest are above 100 Hz, corresponding to the beta dispersion analyzed in Chapter 2. It is observed that the dispersions disappear completely after the four hours, resulting in a flat curve.

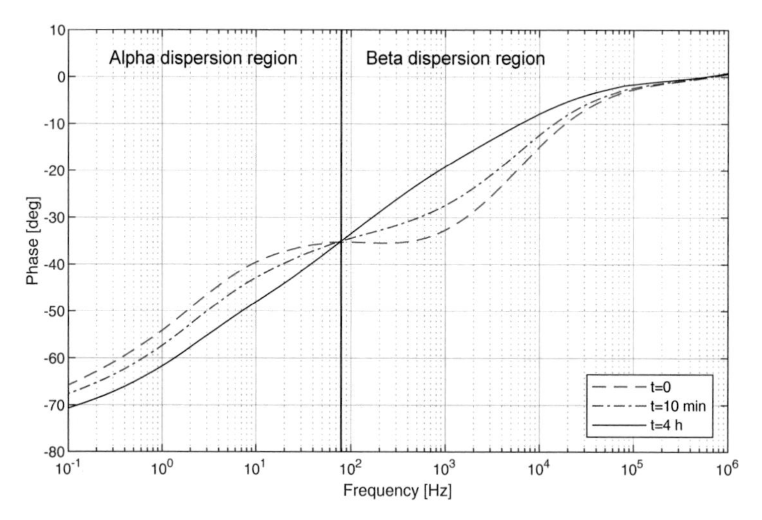

FIGURE 3.11: Impedance phase over the frequency range from 0.1 Hz to 1 MHz.

Conclusion: The impedance of the cell cultures shows two dispersion regions: the alpha and beta dispersions. The alpha dispersion is caused by electrode polarization, explained by the electrochemical double layer. The beta dispersion is originated from the capacitive response of the cellular membranes.

The beta dispersion was found to be disappearing progressively as the experiment takes place. After ten minutes, the impedance was already reduced, and after four hours, the impedance magnitude and phase are considerably flat, without any evidence of the beta dispersion. This is explained by the reduction of membrane capacitance, occurring when cells get damaged. Cell death weakens or even breaks the cellular membrane, and the cytoplasm is then combined with the culture medium. Without membrane-delimited structures, the overall impedance is reduced.

This experiment demonstrated that it is possible to differentiate between viable and non-viable cells using the impedance spectroscopy technique.

3.5 Lactate concentration in athletes

Blood lactate concentration (BLC) is one key parameter to accurately determine the anaerobic threshold of an athlete. The anaerobic threshold is defined as the exercise intensity level where the body cannot dispose the excess of lactate, and therefore it is accumulated in the bloodstream. Lactate accumulation is painful for the athlete, and it is an indicator of the energy efficiency of the body under high load conditions. This measurement is invasive and requires extracting blood samples, to be analyzed in-vitro after the experiments are completed, and therefore it cannot provide instant feedback to the athlete during the sports routine.

Another parameter is the sweat lactate concentration (SLC), which is also measured directly from the sweat. The advantage is that the measurement is non-invasive. However, the difficulties arise from the fact that sweat is a fluid rich in other metabolites, and the initial concentration of lactate varies from one person to another, since lactate is already present in the external layers of the skin even before starting the sports activity. For these reasons, a higher amount of sweat lactate may be measured at early stages of the sports activity. Therefore, a more complex model is required to predict the anaerobic threshold from sweat measurements.

The measurement of BLC and SLC requires external equipment to analyze blood and sweat samples. The instruments allow accurate determination of the lactate and glucose concentrations, but these analysis have to be performed after the sports routine and they do not provide real-time information. This will likely change in the next decades with the incorporation of wearable devices to the consumer market.

Wearable devices include electrodes for real-time recording of biometric signals, but as pointed by Gao in [104], the current technology allows only the measurement of general signals such as the cardiac rhythm and the distance, without any details of the health status at the biological scale. For example, blood and sweat contain different levels of sodium and potassium ions, lactate and glucose. The team of Gao developed a wrist band with integrated functionalized sensors for directly recording the levels of these analytes and electrolytes, and displaying them to the user.

This section includes work in cooperation with the Institute of Sports Medicine from Hamburg University, with the goal of developing a lactate sensor based on impedance spectroscopy. The final objective is to create a device capable of providing instantaneous feedback to athletes, regarding their lactate concentration and anaerobic threshold.

3.5.1 The Krebs and Cori cycles

For all living organisms, energy is stored as glycogen in the liver and skeletal muscle, released to the bloodstream as glucose molecules, and then converted to Adenosine Triphosphate (ATP) which is the basic energy unit necessary for all functions. Every day, the organism processes ATP in quantities similar to the full body weight [105]. For animal organisms there are two mechanisms to synthesize ATP from glucose: the Krebs cycle (aerobic) and the Cori cycle (anaerobic).

Aerobic exercise involves the Krebs cycle, known as the citric acid cycle. This process is more efficient: a single molecule of glucose can yield more than 30 ATP units [106], but it requires more time to complete. This process occurs in the mitochondria and utilizes carbohydrates, fats and proteins to produce glucose. This mechanism does not involve lactate.

For anaerobic exercise, ATP is produced from glucose decomposition in a process explained by the Cori cycle, also known as the lactic acid cycle. This involves two organs: the contracting muscles and the liver. It yields two ATP units per glucose molecule, and lactate is produced and released into the bloodstream. The liver converts lactate back into glucose, requiring six ATP units in the process. Each cycle repetition consumes 4 ATP as the net balance, and therefore this cycle cannot be maintained indefinitely. The Cori cycle is shown in Fig. 3.12.

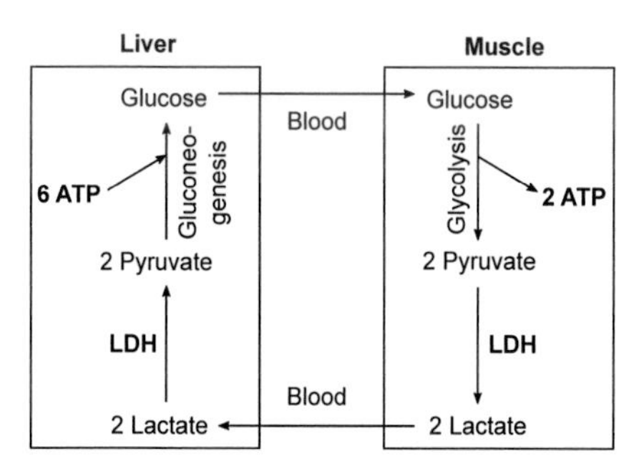

FIGURE 3.12: The Cori cycle for anaerobic exercise.

Both processes have different speeds, and therefore lactate can accumulate if the exercise intensity is elevated above the anaerobic threshold. At this point, the liver cannot process the excess of lactate in the bloodstream and therefore the lactate concentration in blood increases.

3.5.2 In-vitro lactate measurements

A concentrated solution of 60% w/w sodium lactate ($NaC_3H_5O_3$) is prepared by dilution with deionized (DI) water. The sensor is immersed in this solution, and then a signal of 10 mV is applied to the terminals. The impedance is recorded in the range from 1 Hz to 1 MHz. Each measurement is repeated three times, and then the sample is diluted by adding one part of deionized water, lowering the concentration by half. The process is repeated for 20 different concentrations, and the impedance of pure DI water is recorded as reference.

A pinhead connector is used as the sensor, connected to a Gamry Interface 1000 potentiostat using the two-electrode configuration. The impedance magnitude is shown in Fig. 3.13. The impedance is high for pure water (red), and it decreases as the concentration in the sample increases. Conductivity keeps increasing until a saturation point is reached. After this point, if the concentration increases further, the conductivity is slightly reduced.

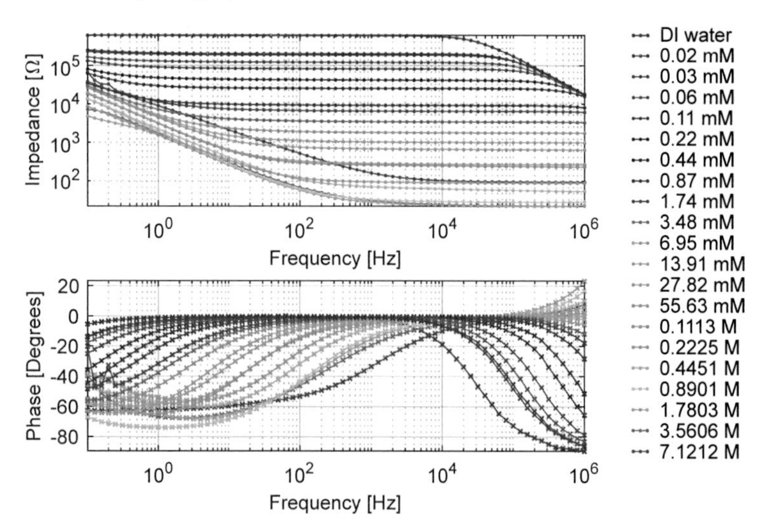

FIGURE 3.13: Impedance of solutions of sodium lactate with concentrations starting from 60% w/w (7.1212 M) and diluted successively by half using DI water.

The data is fitted to the Randles CPE model (see Fig. 2.10) and the solution resistance is plotted against the concentration. The result is seen in Fig. 3.14.

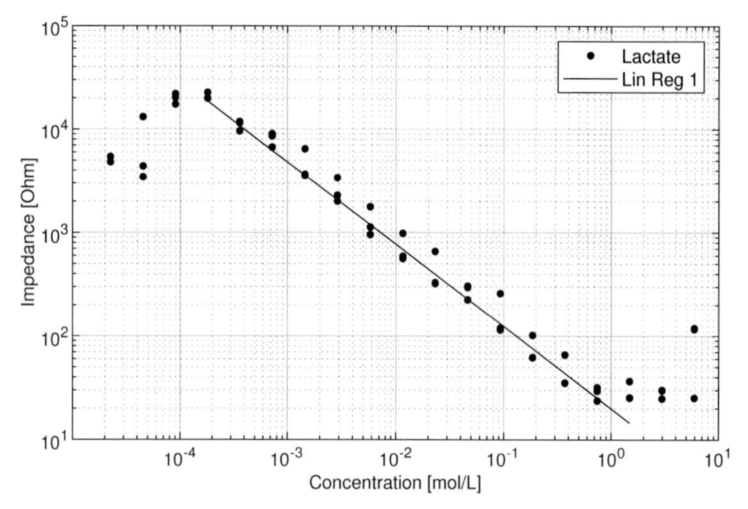

FIGURE 3.14: Impedance Rs of the Randles CPE model, as a function of the concentration, linearized using a log-log plot.

In this figure, the impedance is related directly to the concentration of the sample. This means that the number of molecules can be determined by only looking at the impedance. In this sense, the equation relating the variables is the following:

$$\log(Z) = p_1 \log(C) + p_2 \tag{3.1}$$

where $p_1 = -0.7948$ and $p_2 = 2.9909$. The impedance is calculated as:

$$Z = \exp(p_1 \log(C) + p_2) \tag{3.2}$$

This means the impedance of the solution may be calculated from the concentration, and viceversa:

$$C = \exp\left(\frac{\log(Z) - p_2}{p_1}\right) \tag{3.3}$$

This last equation is of practical importance, because it allows the direct calculation of the concentration of any sample by giving the measured impedance. As a test, consider a sample with an impedance of 1000 Ω in Fig. 3.14. The calculation yields $C = 0.00724$ mol/L which is the concentration of the original sample.

With this method, the concentration of the sample is determined by an instant electronic measurement.

3.5.3 Blood lactate measurements

This section shows how blood lactate varies when athletes perform intensive training. Five persons performed a cycling activity, starting at a low intensity of 50 W, and after five minutes, a linear ramp is produced, increasing the intensity every two minutes until physical exhaustion. The intensity is reduced then to 50 W with a rest time of five minutes, and afterwards another linear ramp is applied untel a second exhaustion point is reached.

Blood samples were taken every three minutes during the exercise routine. The samples were diluted and stored into small vials, and the lactate concentration of each sample was obtained later with enzyme-based sensing equipment.

This experiment is designed in such a way that the lactate increases to two peaks, at the moment of exhaustion. The blood lactate concentration is shown in Fig. 3.15, plotted against the sample number. The average values for all five experiments are displayed in black. The first peak of exhaustion occurs around the sample number 4, and the second peak around the sample number 12.

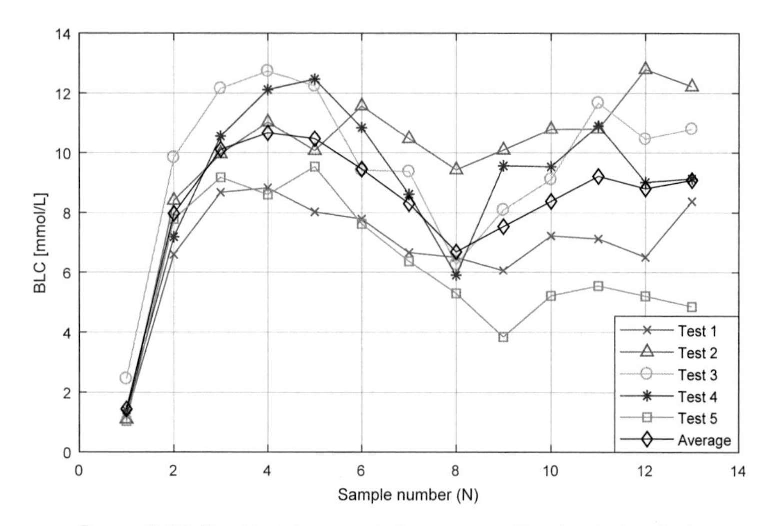

FIGURE 3.15: Blood lactate concentration, measured by chemical method.

This experiment shows how blood lactate is affected by the training intensity. In the next section, an electronic sensor is attached to the skin of the athletes, and the impedance is monitored during the complete sports activity.

3.5.4 Impedance measurements with sensor

In this measurement, two electrodes are placed on the back of the athlete, fixed with a rubber band to keep the electrodes in place and tightly pressed against the body. An Agilent 4284a LCR meter is connected to the two electrodes, to continuously monitor the impedance between them.

The exercise intensity is again adjusted in linear ramps to obtain lactate peaks. The impedance was recorded during the exercise and analyzed with the Randles model with CPE (see Fig. 2.10), and the inverse of the charge-transfer resistance is plotted in Fig. 3.16 for different persons. The blue curves are impedance points, and the red curves are blood lactate values measured by chemical equipment.

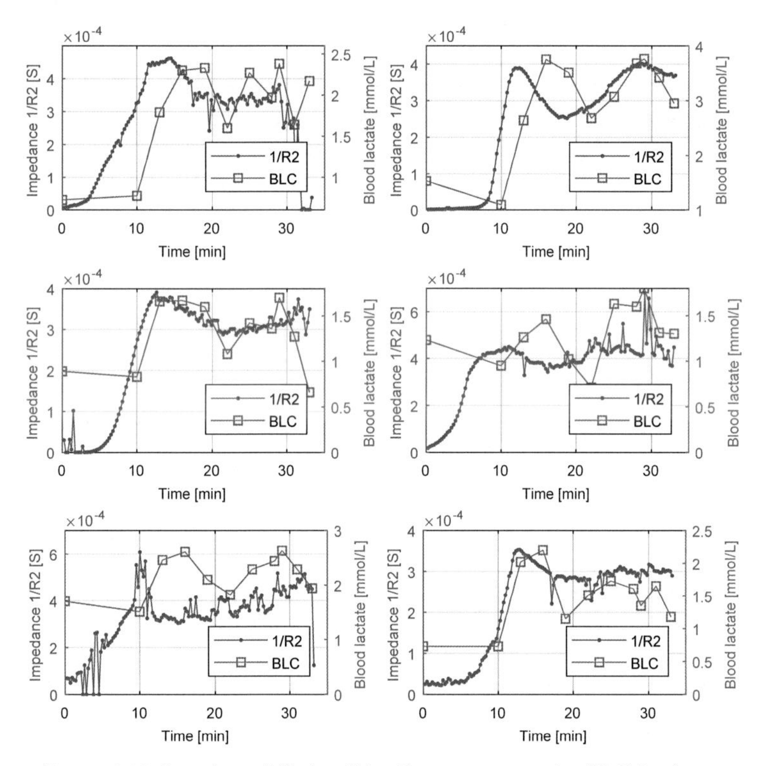

FIGURE 3.16: Impedance (1/Rct) and blood lactate concentration (BLC) for six test subjects.

The increase of the impedance parameter appears slightly in advance, compared with the increase in BLC. Both curves follow the intensity peaks closely, showing that the impedance may be used as an indicator of the anaerobic threshold.

3.6 Chapter conclusion

This chapter shows how impedance spectroscopy is used in different biological and medical applications. In mice, the impedance of cancerous tissue was considerably smaller than the impedance of other healthy tissues. This is the working principle of Electrical Impedance Tomography (EIT). Leg and liver tissues for mice with cancer also exhibited lower impedance than the same tissues on healthy mice, suggesting that the cancer might have already spread to the legs and liver.

In bacterial samples, impedance spectroscopy was able to distinguish between an infected and a healthy sample, proving that it is possible to identify these kind of infections from an electronic measurement without the need of any kind of biological equipment. The impedance was also very different to that of blood samples.

For fruits such as a strawberry, and the food industry in general, the method showed an increase in the impedance during the first days of maturation, and then a strong decline until the fruit is completely ripe. This was in agreement with other experiments in the field. It suggests that the maturation state of food can be observed and monitored electronically.

The next test was the measurement of cell viability, which shown the alpha and beta dispersion regions in Fig. 3.11. The dispersions disappear after four hours, when all the cells are assumed to be dead. This proves that the impedance pattern can be used as an indicator of the cell vitality.

Finally, the experiments regarding lactate opened new possibilities of monitoring the anaerobic threshold. The preliminary tests presented in this section demonstrated the possibility of detecting the lactate concentration *in vitro* using impedance spectroscopy. The plots from Fig. 3.16 show that the impedance from a two-electrode sensor follows the blood lactate concentration (BLC).

These applications can be further explored with the development of a portable impedance spectroscopy system, allowing continuous monitoring in the field. This is the motivation for the remaining parts of this dissertation: the design of portable miniaturized devices for field applications of EIS.

Chapter 4

Impedance measurement circuits

This chapter contains implementation approaches for impedance recording using electronic devices. Existing commercial equipment such as LCR meters and vector analyzers are based on these configurations. The preference of a specific circuit depends on the measurement frequency, as detailed in Fig. 4.1.

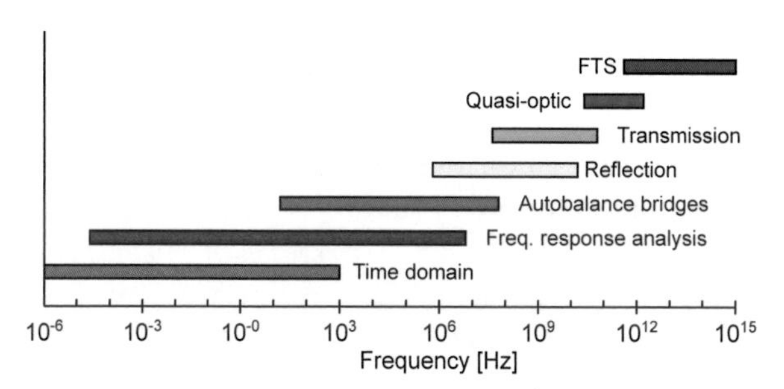

FIGURE 4.1: Impedance measurement techniques by frequency ranges, as used in commercial applications [107].

Time domain spectrometers apply a voltage step to the sample, and calculate the capacitance and resistance directly, by measuring the time it takes to charge and discharge, which is proportional to τ = RC.

Frequency response analysis (FRA) use a lock-in amplifier to generate a signal, and a multiplier to detect the signal after the sample. The output voltage is filtered and processed to recover the magnitude and phase information.

Auto-balancing bridges measure the current flowing through the sample by using a current-to-voltage converter. This is a standard technique used in commercial LCR meters under 30 MHz.

Reflection and transmission methods are based on the properties of a two-port network. If a coaxial cable is used for the measurement, any discontinuities in the cable will introduce reflections. The impedance can be calculated from the reflection coefficient Γ. These methods can achieve frequencies up to 40 GHz.

Quasi-optic methods (up to 1 THz) and Fourier-transform spectrometers (FTS, up to 1400 THz) are very specific constructions that apply electromagnetic waves to the sample and deflect it with mirrors, in a similar way to a Mach–Zehnder interferometer. The deviations of the beam compared with itself give enough information to determine the permittivity.

In the following sections, some circuit topologies that may be integrated into an ASIC are discussed.

4.1 Wheatstone bridge

The Wheatstone bridge is a circuit capable of measuring the impedance of an unknown resistor with high accuracy [108]. It is composed of four resistors and a galvanometer, as shown in Fig. 4.2. The resistors R_1, R_2 and R_3 are known, whereas the impedance Rx is unknown.

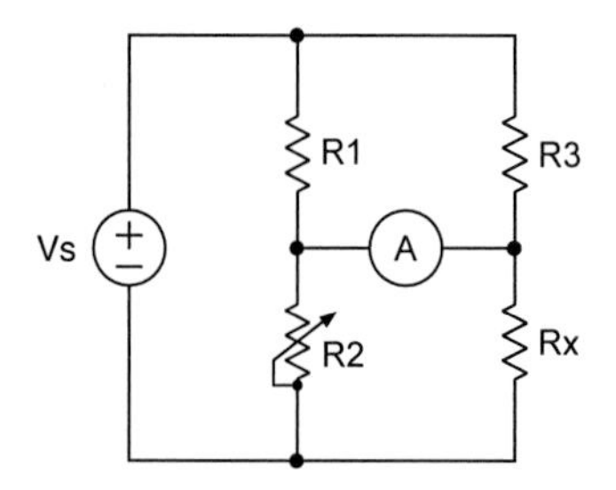

FIGURE 4.2: Schematic of the Wheatstone bridge.

The resistor R_2 is adjusted until the circuit is properly balanced, and therefore the current flow through the galvanometer is zero. This method is highly accurate since any slight deviation from zero current can be detected by the galvanometer.

At the equilibrium condition, the impedance ratio of the two branches is constant, and the unknown impedance can be calculated in terms of the other three resistors as:

$$\frac{R_2}{R_1} = \frac{R_X}{R_3} \Rightarrow R_X = \frac{R_2}{R_1} \cdot R_3 \tag{4.1}$$

Additionally, if the resistor R_2 cannot be adjusted, the current flowing through the galvanometer can be used to calculate the impedance, using the Kirchhoff's law. However, this limits the range in which the resistor R_X can vary, and the sensitivity of the measurement is decreased.

This circuit is useful to measure resistive loads only, but it is given here due to its simplicity. More advanced methods actually used in commercial devices from the market are shown below.

4.2 Auto-balancing bridge method

The impedance of a device-under-test can be determined by sampling the voltage and current simultaneously. This approach is based on the current-to-voltage converting configuration of operational amplifiers. The general schematic of this configuration is shown in Fig. 4.3.

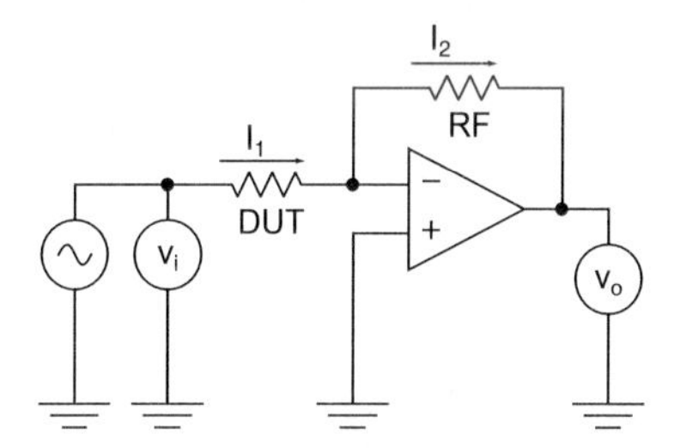

FIGURE 4.3: Schematic of the auto-balancing bridge method

In this circuit, the node after the DUT is called a virtual ground. The current through the sample is calculated as $I_1 = v_i/Z_{DUT}$ and this current is the same as the current I_2, since no current flows through the non-inverting terminal of the amplifier. This produces a voltage drop $v = I_2 R_f$ across the feedback resistance, which appears inverted at the output as $v_o = -I_2 R_f$.

By considering the equality $I_1 = I_2$ it may be written:

$$\frac{v_i}{Z_{DUT}} = \frac{-v_o}{R_f} \tag{4.2}$$

The impedance of the DUT is calculated then as:

$$Z_{DUT} = -R_f \cdot \frac{v_i}{v_o} \tag{4.3}$$

The input and output voltages are sampled by analog-to-digital converters (ADC) and processed using the Fast Fourier Transform (FFT) to remove the noise and calculate the magnitude and phase of the signals. The impedance magnitude is obtained by substituting the amplitudes $|v_i|$ and $|v_o|$ in the previous equation. The impedance phase is equal to the phase angle between v_i and v_o.

The FFT is an algorithm that computes the frequency components of a signal, decomposing it in terms of sinusoidal oscillations. The result is a graph known as the Power Spectral Density (PSD), including all frequency components with their corresponding amplitude and phase. By searching this PSD and locating the fundamental frequency, one can obtain the amplitude and phase of the main frequency component of the signal. The DC offset is also extracted. This has an additional advantage of removing all the noise from other frequencies, focusing only on the signal of interest.

To compute the FFT from digital signals sampled by ADCs, a memory with a specific amount of positions is required. Usually, powers of 2 are used for simplicity and to improve the speed and efficiency of the algorithms. The FFT is calculated usually in blocks of 256, 512 or 1024 elements, always in multiples of 2. One of the most widely used algorithms is the Discrete Fourier Transform (DFT).

For example, a signal can be sampled in the time domain by using a sample frequency f_s, and 512 samples are obtained and stored. These samples are processed in a computer or an embedded system, and the output is a PSD with 512 frequency components, ranging from 0 to $f_s/2$.

This measurement approach is used in Chapter 5 and Chapter 6 for the design of portable and miniaturized impedance measurement devices.

4.3 Frequency response analysis method

The frequency response analysis (FRA) method [109, 110] is shown in Fig. 4.4. This uses a quadrature oscillator, two analog mixers and two integrators. The circuit configuration is very similar to a lock-in amplifier. The DC output voltages are proportional to the real and imaginary parts of the admittance of the DUT.

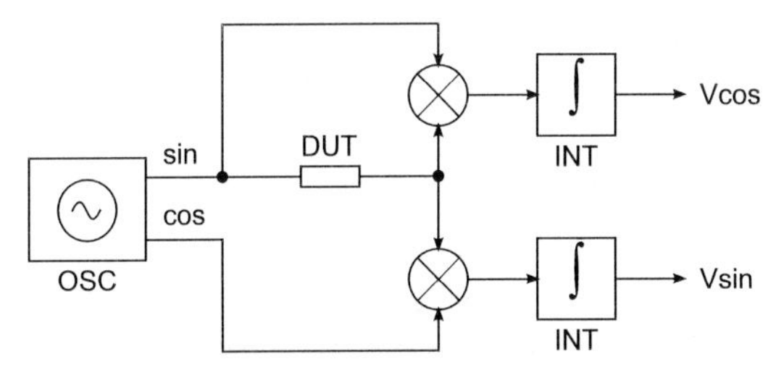

FIGURE 4.4: Frequency response analysis method (FRA [109]).

The output signals are calculated by applying trigonometric identities, as follows:

$$\cos(\alpha) \times \cos(\beta) = \frac{1}{2}\cos(\alpha-\beta) + \cos(\alpha+\beta) \tag{4.4}$$

$$\sin(\alpha) \times \cos(\beta) = \frac{1}{2}\sin(\alpha+\beta) + \sin(\alpha-\beta) \tag{4.5}$$

Therefore:

$$\begin{aligned} \mathrm{Vcos} &= I_0 \sin(\omega t) \times \sin(\omega t + \phi) && (4.6)\\ &= \frac{I_0}{2}[\cos(\phi) - \cos(2\omega t + \phi)] && (4.7)\\ &\approx \frac{I_0}{2}\cos(\phi) && (4.8) \end{aligned}$$

$$\begin{aligned} \mathrm{Vsin} &= I_0 \cos(\omega t) \times \sin(\omega t + \phi) && (4.9)\\ &= \frac{I_0}{2}[\sin(\phi) + \sin(2\omega t + \phi)] && (4.10)\\ &\approx \frac{I_0}{2}\sin(\phi) && (4.11) \end{aligned}$$

The terms involving $2\omega t$ in eq. (4.7) and eq. (4.10) are filtered by the integrators, which are low-pass filters. The oscillator current is then calculated as:

$$I_0 = 2\sqrt{V\cos^2 + V\sin^2} \tag{4.12}$$

If the oscillator voltage is known, the impedance can be computed by Ohm's law.

This is the measurement technique of the ASIC 2 and ASIC 3, which is explained in detail in Chapter 7 and Chapter 8.

4.4 Network analysis method

At higher frequencies, instead of sampling individual voltages or currents, the power of a signal transmitted or reflected can be measured with higher accuracy.

Consider the circuit of Fig. 4.5, where Z_0 is the characteristic impedance of a transmission line (for example a coaxial cable), and Z is the impedance of the termination, or in this case the device-under-test:

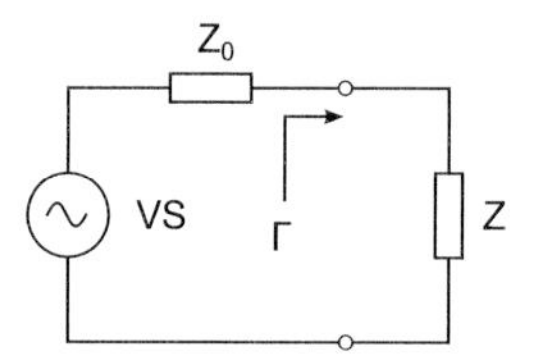

FIGURE 4.5: Schematic of the network analysis method.

The impedance Z can be calculated from the reflection coefficient Γ as described by the following equations:

$$\Gamma = \frac{Z - Z_0}{Z + Z_0} \tag{4.13}$$

where Z_0 is the characteristic impedance of the transmission line.

The reflection coefficient has a value of -1 if the impedance is a short-circuit, +1 for an open-circuit and 0 for a perfectly matched impedance.

This method cannot be directly implemented on an ASIC due to the complexity of power detection circuitry in modern equipment. However, vector network analyzers (VNA) may be used in laboratories for the characterization of impedances in frequency ranges extending over several hundreds of gigahertz. Modern VNAs may achieve accurate sensing over 100 GHz.

The impedance can be calculated also from the standing wave ratio (SWR):

$$\text{VSWR} = \frac{1 + |\Gamma|}{1 - |\Gamma|} \tag{4.14}$$

4.5 Chapter conclusion

Most of the commercial impedance measurement devices such as LCR meters are based on circuit topologies such as the auto-balancing bridge method (ABBM) and the frequency response analysis (FRA).

In addition to the topologies presented in this chapter, there are many other circuit topologies for the determination of impedance. Digital circuits such as the AFE4300 from Texas Instruments, and the Impedance-to-Digital Converters such as the AD5933 from Analog Devices, are used with the same purpose. These are commercial integrated circuits that may serve to build portable platforms.

The ABBM is used in Chapter 5 for the development of a portable impedance spectrometer, using standard commercial components.

The next step in miniaturization is achieved by using full-custom ASICs. This is required for applications where a completely optimized device is needed, for example in implants or wearable technology. In this dissertation, the ABBM topology is used in the ASIC 1, and the FRA topology is the base of the ASIC 2 and ASIC 3, where the high performance of current microelectronic devices can be used on silicon to extend the frequency ranges of these techniques to the gigahertz range.

Chapter 5

Demonstrator of a portable impedance spectrometer

The portability of commercial impedance spectrometers is limited due to the large size of the equipment. Bench LCR meters and potentiostats require a large space in the laboratory, and they need to be constantly powered from the AC line. In addition these devices are costly and require trained personal to be operated.

Medical applications require a fast and autonomous way to take impedance measurements with a single touch of a button, since experiments are often performed by doctors and scientists outside the engineering field. One example is the development of wearable devices that can record the impedance during intense sports activity, which is studied in Chapter 3.

This chapter describes the design and implementation of a portable impedance spectroscopy device, by implementing the Auto-Balancing Bridge Method on a PCB using commercially available parts such as operational amplifiers and resistors, with the objective of creating a portable and standalone measurement device, which may be operated by persons outside the engineering field.

A first prototype is working in the frequency range from DC up to 125 kHz. The device is used to measure the impedance of yeast cell cultures, and it is compared against the impedance from a commercial LCR meter.

5.1 Block diagram

The design of the first prototype was carried out by Adolfo Fernandez, and the results are published in [47]. This device allows measurement of impedances up to 125 kHz, with an accuracy of 5% at 40 kHz, and of 15% at 100 kHz.

The general block diagram of the prototype is presented in Figure 5.1.

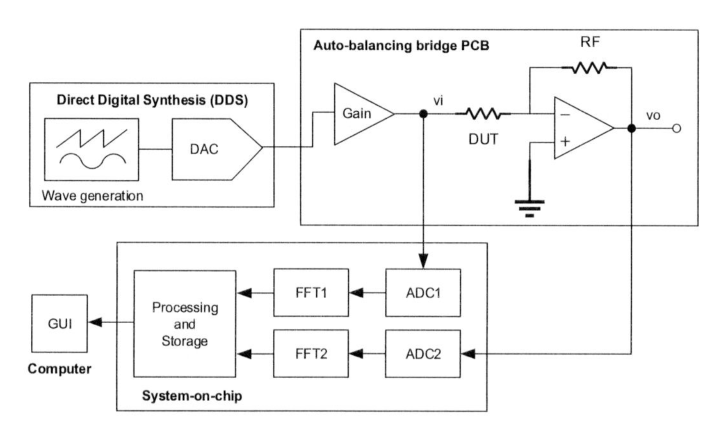

FIGURE 5.1: Topology of the potentiostat based on the ABBM method.

The waveforms for stimulation are generated from a Direct Digital Synthesis (DDS) integrated circuit, which is a digital component that includes a Read-Only Memory (ROM) with preloaded information about different waveforms such as a sine, triangular and square wave, and a Digital-to-Analog Converter (DAC).

The impedance is computed on a System-on-Chip (SoC) device, which includes ADCs for sampling the input and output voltages of the ABBM, and then processes the waveforms using the Fast Fourier Transform (FFT) for filtering and calculation of the magnitude and phase of the signals. The impedance data is then stored in a local SD card and may be observed graphically during the measurements using a VNC connection through Ethernet.

The selected SoC board is the Terasic DE0-nano-SoC, which is based on the Altera Cyclone V. It includes an FPGA with 40K logic elements, and an ARM dual-core processor running at 925 MHz. The board runs a custom version of Ubuntu Linux.

5.2 Design of the analog circuitry

The schematic of the designed PCB is shown in Figure 5.2. This board includes the generation of an analog signal, amplification, filtering and the stimulation of the device-under-test, and the measurement of the stimulation current by the auto-balancing bridge method.

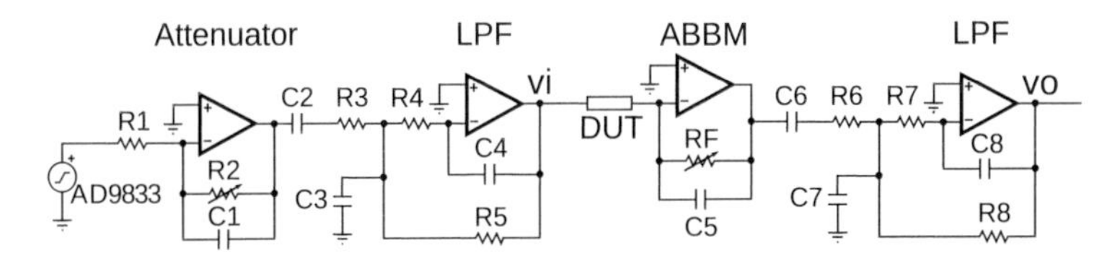

FIGURE 5.2: Schematic of the analog circuitry, consisting of an attenuator, a current-to-voltage converter (ABBM method) and two low-pass filters.

The AD9833 is a programmable waveform generator, operating under the concept of Direct Digital Synthesis (DDS). It generates sinusoidal, triangular and square signals up to 12.5 MHz, with a constant output voltage of approximately 600 mV and a DC offset of 1.65 V.

This voltage of 600 mV is nominally too large for stimulation of biological tissues and cells, since the cellular membranes are thin and can break with elevated electric fields. Therefore, the use of an operational amplifier as an attenuator was implemented. The attenuation adjusts the generated signal to a range of 0-200 mV by using a potentiometer.

The DC offset is eliminated by using a 10 µF capacitor (C2) at the output of the attenuator. This is done to prevent damages to the cells, and also with the intention of setting the operating point of the op-amps at ground, since they are powered by dual ±3.3 V power supplies.

The second-order low-pass filters are based on the multiple-feedback topology, and they are designed to have a cut-off frequency of 16 MHz, to filter the noise generated by the 25 MHz clock used by the waveform generator.

Finally, the input (v_i) and output (v_o) voltages are sampled by the on-board ADCs of the development kit, with a maximum sampling frequency of 250 ksps.

5.3 Design of the digital logic

The device requires digital logic to perform automatic impedance spectroscopy curves. This is achieved by programming a Python script which runs on the embedded Linux system. The script performs the following three functions:

1. Programming the waveform generator via I2C, by setting the frequency, amplitude, and signal type (triangular, square or sinusoidal).
2. Measurement of the input and output voltages by taking 1024 samples using the ADCs, and
3. Computing the FFT and the impedance, storing the result on the SD card.

The process is repeated once for each frequency point, to perform the sweep in the range from DC up to 125 kHz, in steps of approximately 100 Hz.

5.4 Frequency response measurements

The frequency response of the assembled circuit was obtained by applying an input signal from a waveform generator, and measuring the voltage at the output of the second low-pass filter.

The curve on Fig. 5.3 show a measured cut-off frequency of 5 MHz which is enough for the impedance measurements in the present frequency range.

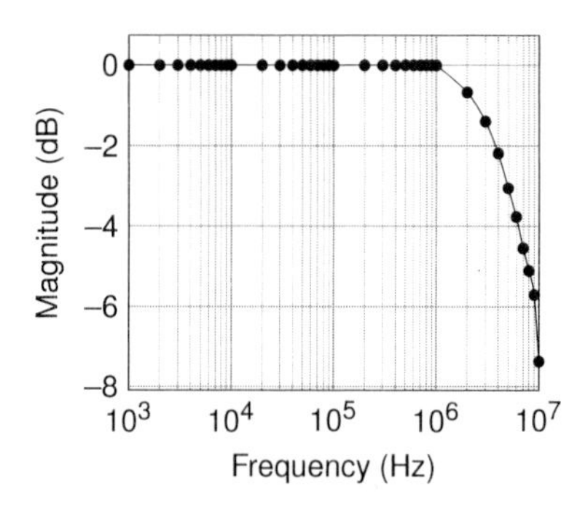

FIGURE 5.3: Measured Bode plot after the second low-pass filter.

5.5 Experiments with constant known loads

The impedance of constant RC-series loads was measured to demonstrate the accuracy of the device. A resistor of 10 kΩ is connected in series with different capacitors as in Table 5.1 using a stimulation frequency of 40 kHz.

TABLE 5.1: Measurement of an RC series circuit with R=10 kΩ at 40 kHz.

C (pF)	Xc (Ω)	Theoretical		Measured		Error %	
		|Z|	θ	|Z|	θ	|Z|	θ
10	397887.36	398013.00	-88.56	387550.95	-84.47	2.63	4.62
47	84656.88	85245.46	-83.26	85452.26	-82.67	0.24	0.71
100	39788.74	41026.13	-75.89	39623.77	-74.67	3.42	1.61
220	18085.79	20666.30	-61.06	21542.85	-60.11	4.24	1.55
470	8465.69	13102.21	-40.25	12987.48	-38.93	0.88	3.27
2100	1894.70	10177.91	-10.73	10250.31	-10.66	0.71	0.67

It is observed that the accuracy of the measurement is of at least 5% for the measurements at 40 kHz. Both magnitude and phase can be computed within the same level of precision.

The measurement of a 3 kΩ resistor is presented in Fig. 5.4, using the full frequency sweep range from 0 up to 125 kHz. The expected impedance is given as a straight horizontal line. As it can be seen, the absolute error in this plot is at most of approximately 300 Ω at frequencies above 100 kHz, which corresponds to a 10% of the measurement.

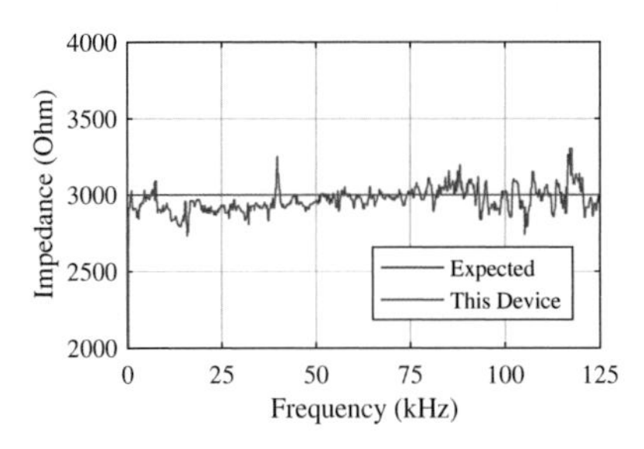

FIGURE 5.4: Magnitude of a 3 kΩ resistance, measured from DC up to 125 kHz.

5.6 Experiments with yeast

The designed measurement device was used to observe the impedance of yeast cell cultures as they grow in a plastic chamber with stainless steel electrodes. The chamber was designed to have a volume of one cubic centimeter, with parallel-plate electrodes for stimulation with an uniform electric field.

For this experiment, the container was filled with yeast until half of it, and the rest of the chamber was filled with tap water to the top. Then, the board kept sampling the impedance over 30 minutes, taking one sample per second. The stimulation voltage was 100 mVpp and the frequency is of 40 kHz.

The experiment is shown in Fig. 5.5.

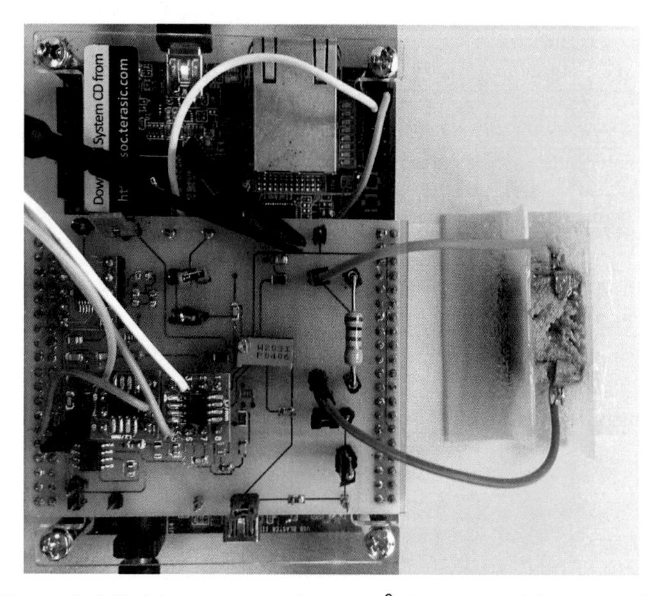

FIGURE 5.5: Prototype 1 connected to a 1 cm^3 chamber containing yeast cells.

The experiment was repeated with an Agilent 4984A LCR meter under the same experimental conditions, and both results are plotted in Fig. 5.6.

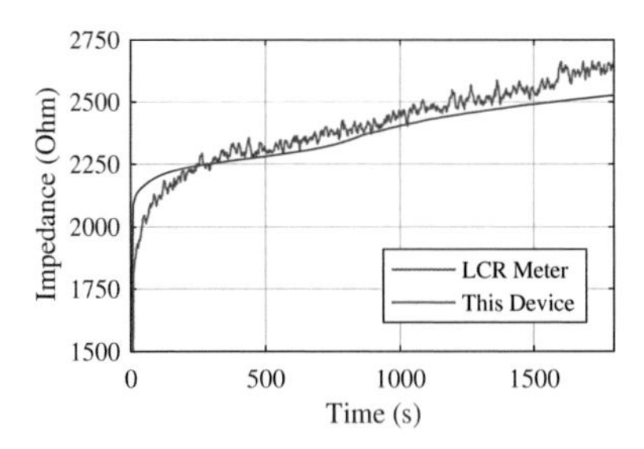

FIGURE 5.6: Impedance spectroscopy of the growth of yeast cultures at 40 kHz, at room temperature, using a 1 cm^3 chamber with stainless steel electrodes.

5.7 Chapter conclusion

This chapter showed the design and implementation of an impedance spectroscopy measurement device, which is portable and operates in a standalone way. The assembled demonstrator is portable in the sense that it is totally autonomous, does not require an external computer to operate, and its dimensions are considerably reduced. The demonstrator measures 10x7x5 cm and can be carried together with a 9 V battery for fully autonomous operation. The total cost of the assembled device, including the system-on-chip evaluation board, is of approximately 160 €.

In contrast, commercial bench LCR meters have dimensions in the range of 40x40x20 cm, require a permanent power supply of 230 V, and need a computer for sending command sequences for impedance spectroscopy, for example for performing frequency sweeps or continuous time sampling. It can be appreciated that the demonstrator offers a significant improvement in portability, cost, power consumption and autonomy of operation.

The portable device is used to measure the impedance of RC loads with an accuracy of 5% at 40 kHz, as observed in Table 5.1. The device also allows to perform frequency sweep measurements from 0 to 125 kHz as shown in Fig. 5.4, and continuous impedance monitoring as presented in Fig. 5.6. These capabilities are typical of standard potentiostats, and together with the toolbox developed in Chapter 2, the demonstrator allows the measurement and processing of electrochemical impedance spectroscopy data, together with the fitting software for equivalent circuit models. This constitutes a fully portable EIS solution.

The impedance of yeast cell cultures was monitored for 30 minutes using the demonstrator, and the results are comparable to the measurements from a commercial LCR meter. It was observed that the impedance of the cell culture was increasing, giving evidence of the cell growth.

Further extension of the measurement frequencies is possible by replacing the ADCs with high-speed components, which would allow sampling the input and output voltages at 25 MSPS, which is one hundred times faster than the present demonstrator. This could extend the frequency range up to 12.5 MHz.

It is demonstrated that the ABBM method works, and that it is possible to integrate it into an ASIC, since all the components used in the design can be manufactured within the power and space limitations of a custom integrated circuit. This is the topic of the next chapter.

Chapter 6

Miniaturized portable impedance measurement system

This chapter covers the specifications, design and testing of an impedance spectroscopy ASIC under 40 kHz. The chip was designed by J. Tomasik, W. Galjan and C. Hafkemeyer, and produced in the AMS H35B4 350 nm technology. The results from this chapter are published in [46]. The ASIC is designed to measure the impedance of a device under test (DUT) with magnitude and phase, in the frequency range from DC up to 40 kHz. The device is pictured in Fig. 6.1. The design includes an internal oscillator (OSC), two low-pass filters (LPF), a current-to-voltage converter (I2V), two analog-to-digital converters (ADC) and a serial register (SR32) for storing the configuration of each block.

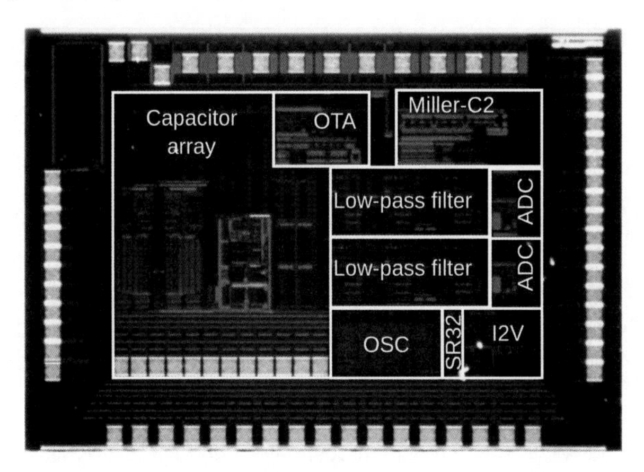

FIGURE 6.1: Photography of the ASIC 1 showing the functional blocks [46].

6.1 Chip architecture

The ASIC 1 is an implementation of the Auto-Balancing Bridge Method (ABBM) [111]. The block diagram of the ASIC is presented in Fig. 6.2.

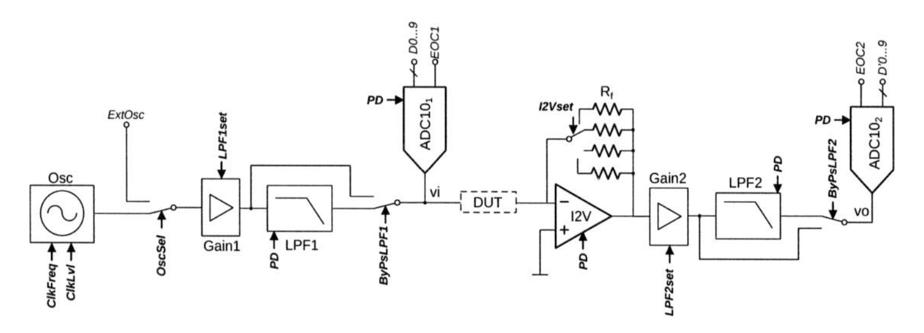

FIGURE 6.2: Architecture of the ASIC 1.

The operation of the ASIC is as follows. The stimulation waveform can be selected from the internal oscillator or an external source. This signal is amplified and filtered by the LPF 1. At this point marked as v_i, the signal is sampled by an ADC and then applied to the DUT. The current passes through the current-to-voltage converter, which produces an inverted output voltage proportional to the current. This signal is again amplified and filtered by the LPF 2, before being sampled by a second ADC at the point labeled as v_o.

Each block of this diagram will be explained in the following paragraphs.

6.1.1 Configuration register

The ASIC 1 stores its configuration in a 32-bit serial input, parallel output (SIPO) register. This allows to adjust the oscillator frequency and amplitude, set the low-pass filter gains, select the feedback resistor and optionally bypass the low-pass filters. The programming of the register is achieved with a MSP430 microcontroller from Texas Instruments.

By default, the register is completely erased when the chip is powered on. The default configuration sets the chip to operate with the internal oscillator, with the low-pass filters enabled, the gain of each LPF set to Av=1 and the feedback resistance of the I2V converter set to 1 MΩ.

6.1.2 Internal oscillator

The generation of a sine wave can be achieved by two mechanisms: analog oscillators and digital waveform generators. Tuning an analog oscillator requires varying the capacitance of the LC resonant circuit, which makes tuning a manual process. Digital waveform generators have the advantage of being programmable, allowing to change the frequency and amplitude during chip operation.

Obtaining a pure sine wave directly from digital calculations is computationally intensive, and therefore most chips include a look-up table with the sinusoidal function stored in a read-only memory (ROM). The table is fed to a digital-to-analog converter (DAC) to produce the analog output, and the speed of the memory counter defines the output frequency. This approach requires a large memory, which is a disadvantage for ASIC designs since the chip area is limited.

In the ASIC 1, the on-board oscillator was designed keeping in mind an area-efficient approach. This oscillator is based on the oversampling D/A conversion technique [112]. The idea is to build a digital resonator circuit, and use a low-pass filter to obtain the sine waveform at the output, without the need of a DAC. The area cost of this generator is of 4 multi-bit adders, 4 registers, a 2-input multiplexer and an analog low-pass filter.

The oscillator requires an input square waveform of 12.8 MHz and 3.3Vpp to generate a sinusoidal signal of variable amplitude and frequency. The oscillation frequency and amplitude can be configured in the main register, according to Table 6.1.

TABLE 6.1: Configuration bits for the oscillator.

Bits	Name	Description
Reg<1:7>	ClkFreq	0010001: 39 kHz
		1010110: 15 kHz
		0011000: 3 kHz
Reg<8:14>	ClkLvl	0000001: 1 V
		0010100: 500 mV
		0010000: 100 mV
		1000000: 25 mV
Reg<15>	OscSel	0: Internal
		1: External

6.1.3 Low-pass filters

The low-pass filters used in the ASIC are fourth-order LPFs, which are designed by cascading two second-order filters in the Sallen-Key configuration. Each filter includes a preamplifier with configurable gain. The schematic of a LPF is shown in Fig. 6.3.

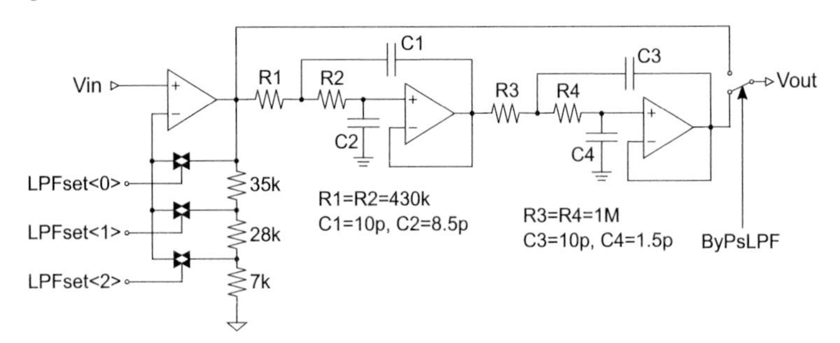

FIGURE 6.3: Schematic of a low-pass filter.

The Sallen-Key filters have unity gain and the design parameters are the cut-off frequency f_c, the natural frequency f_0, the quality factor Q, the attenuation α and the damping ratio ζ, which are explained as follows.

The cut-off frequency of each Sallen-Key filter is given by:

$$f_c = \frac{1}{2\pi\sqrt{R_1C_1R_2C_2}} \tag{6.1}$$

The transfer function for one Sallen-Key filter is given by:

$$H(s) = \frac{\omega_0^2}{s^2 + 2\alpha s + \omega_0^2} \tag{6.2}$$

where $\omega_0 = 2\pi f_0 = 1/\sqrt{R_1C_1R_2C_2}$ is the natural angular frequency, and

$\alpha = \zeta\omega_0 = \frac{\omega_0}{2Q} = \frac{1}{2C_1}\left(\frac{1}{R_1} + \frac{1}{R_2}\right)$ is the attenuation. Therefore,

$Q = \frac{\omega_0}{2\alpha} = \frac{\sqrt{R_1R_2C_1C_2}}{C_2(R_1 + R_2)}$ is the quality factor.

From the previous equations, both filters have a cut-off frequency of 40 kHz. The Q factor is 0.542 for the first filter, and 1.291 for the second.

The gain of the preamplifier is set using the configuration register. In addition, the filter can be disabled by configuring the ByPsLPF bit of the register. The configuration options for the LPF are described in Table 6.2.

TABLE 6.2: Configuration bits for the low-pass filters.

Bits	Name	Description
Reg<16>	ByPsLPF1	0: LPF1 Enabled
		1: LPF1 Disabled
Reg<17>	ByPsLPF2	0: LPF2 Enabled
		1: LPF2 Disabled
Reg<18:20>	LPF1set	Gain Adjustment 1
		000: Av = 1
		110: Av = 2
		111: Av = 6
		101: Av = 10
Reg<21:23>	LPF2set	Gain Adjustment 2
		000: Av = 1
		110: Av = 2
		111: Av = 6
		101: Av = 10

6.1.4 Analog-to-digital converters

The voltage applied to the DUT (v_i) and the output voltage (v_o) are sampled by using analog-to-digital converters, which are used to interface the ASIC to the external processing circuitry. The impedance computation is done in a computer or an embedded system with DSP capabilities.

The ADCs used in the design are the ADC10A standard cells from the AMS analog standard cell library, which are based on the successive approximation method (SAR-ADC). These devices have a 10-bit output, providing a resolution of 3.2 mV for a supply voltage of 3.3 V.

The ADCs require only a clock signal and a start signal, and they include an end-of-conversion (EOC) flag, which is asserted when the data is available at the digital output terminals.

6.1.5 Current-to-voltage converter

The main block of the ABBM topology is the current-to-voltage converter (I2V), which is a transimpedance amplifier. This is implemented by a low-noise operational amplifier from the analog standard cell library, the OP_LN, which is the same amplifier used in the LPFs.

The schematic of this block is shown in Fig. 6.4. The feedback resistors are integrated on-chip and can be adjusted to four different values: 10 kΩ, 100 kΩ, 1 MΩ and 10 MΩ depending on the expected DUT impedance.

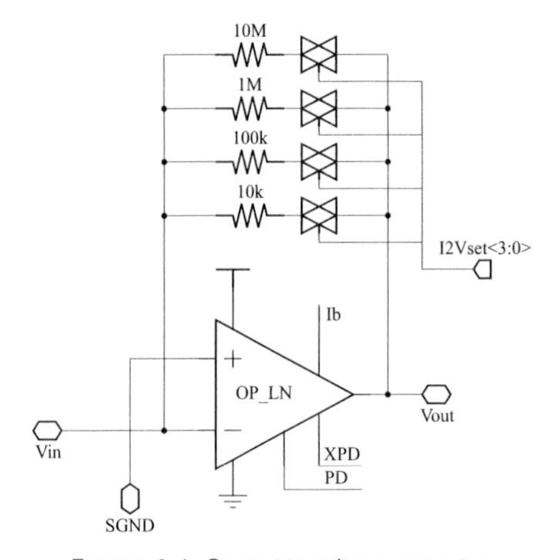

FIGURE 6.4: Current to voltage converter.

The feedback resistors are used to set the gain of the amplifier, in combination with the DUT. The gain of the circuit is defined as $A = -R_f/Z_{DUT}$.

For a reliable operation, the gain should be less than 1 in order to avoid saturating the output of the chip. At the same time, the gain should be high enough to ensure that the output voltage can be measured within the ADC resolution.

For example, if the selected feedback resistance is Rf = 10 kΩ, the expected impedance for the DUT should be larger than 10 kΩ. This condition gives four recommended operation regions: A, B, C and D, described in Table 6.3.

TABLE 6.3: Operation ranges for the ASIC 1.

Region	Reg<24:27>	Rf	Expected DUT range
A	1010	10 kΩ	10 kΩ – 100 kΩ
B	0110	100 kΩ	100 kΩ – 1 MΩ
C	0000	1 MΩ	1 MΩ – 10 MΩ
D	0011	10 MΩ	10 MΩ – inf

To give an example of the operation ranges and how they should be used, consider the following simulation. Here, the circuit is used to measure the impedance of a load in the full range, by sweeping the DUT from 10 kΩ to 100 MΩ. The results of this simulation are presented in Fig. 6.5.

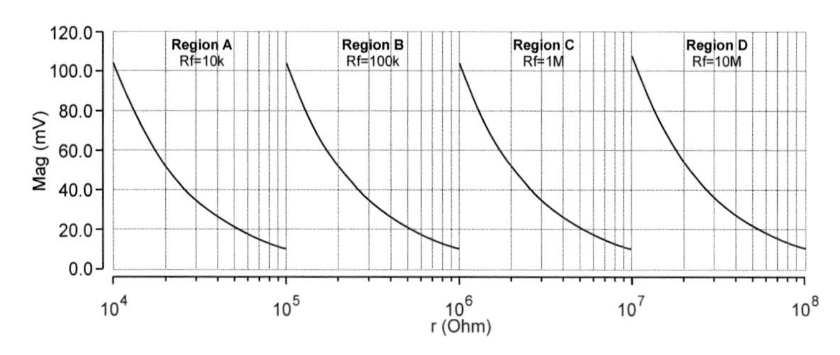

FIGURE 6.5: Simulation of the output voltage while sweeping the impedance of the DUT from 10 kΩ to 100 MΩ.

In this simulation, an external oscillation of 100 mVpp and 20 kHz is applied to the input of the circuit. The gain of the LPFs is set to 1. The first impedance point is DUT=10 kΩ, which for Rf=10 kΩ would give unity gain, producing 100 mV at the output.

The increase of the DUT modifies the gain of the I2V converter, decreasing it as the impedance of the DUT increases. Therefore, the amplitude of the output voltage decays as the resistance is increased.

When the impedance of the DUT gets to 100 kΩ, the output voltage is too small since the gain of the I2V is 0.1, and then the next feedback resistance is used to bring back the gain to 1.

6.2 Impedance calculation example

This section illustrates the measurement of a 100 kΩ resistor with the circuit.

The internal oscillator produces a voltage of 100 mVpp. This signal is shown in Channel 1 of the oscilloscope diagram from Fig. 6.6. The oscillator is amplified by the first LPF, with a gain of 6. Therefore, a signal of 600 mVpp appears at Channel 2. This signal is applied to the DUT. The signal from Channel 3 corresponds to the virtual ground node, at the inverting terminal of the operational amplifier, which holds only a DC level of 1.65 V corresponding to the signal ground. Finally, the output voltage appears in Channel 4. The feedback resistor was set to 10 kΩ.

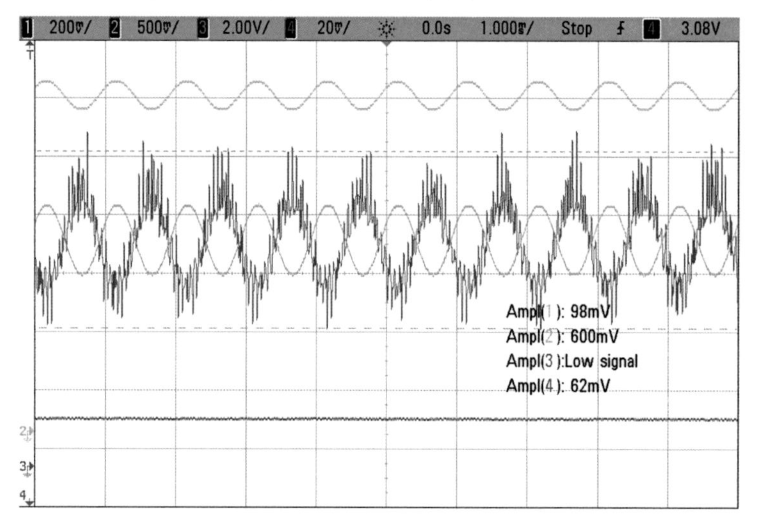

FIGURE 6.6: Waveforms for the measurement of a resistor, showing v_s = 98mV, v_i = 600mV and v_o = 68mV.

With this information, the impedance and phase can be calculated.

The impedance of the DUT is $Z_{DUT} = -R_f \frac{v_i}{v_o} = -10k\Omega \cdot \frac{600mV}{62mV} = 96.77k\Omega$.

The error percentage is $\epsilon = (100k - 96.77k)/(100k) = 3.23\%$.

The phase of the DUT corresponds to the phase difference between v_i and v_o. In this case, a phase difference of 180 degrees is visible in the diagram, which means the phase angle is 0 since the I2V inverts the signal. The exact determination of this angle is possible by using the FFT, which will be addressed in the next section.

6.3 LabVIEW Implementation

Impedance monitoring of biological cell cultures with the ASIC is only possible if the impedance calculation is performed automatically and in real-time during the experiments. To achieve this, the experimental setup of Fig. 6.7 was assembled.

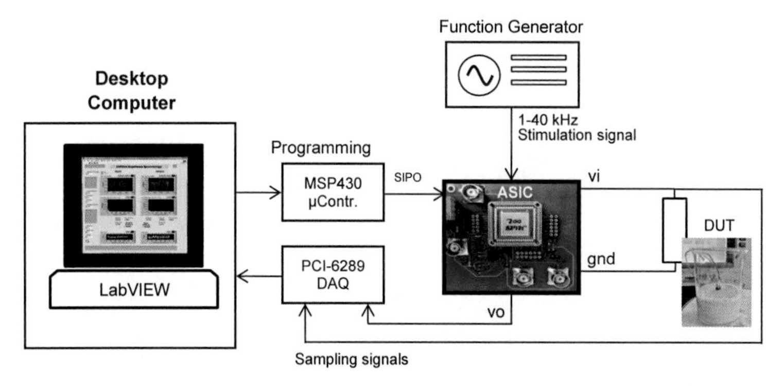

FIGURE 6.7: Experimental setup for impedance spectroscopy with LabVIEW.

The setup uses a desktop computer for writing the configuration to the ASIC through the MSP430 controller, allowing to write any combination of programming options in the internal configuration register.

The external function generator allows to finely tune the amplitude and frequency of the stimulation voltage, beyond the predefined values from the chip oscillator. A sine wave with DC offset of 1,65 V should be used. The amplitude for stimulation can be set to any value between 0 and 1.65 Vp to remain within the 3.3 V rail-to-rail operation.

The ASIC voltages are sampled by using a National Instruments PCI-6289 DAQ card, at a minimum sampling frequency of 80 kHz per channel. The data is transferred to the computer, where a LabVIEW application is running to compute the impedance and store it on the hard disk.

To calculate the impedance of the DUT, the discrete Fourier transform (DFT) is used to obtain the frequency spectrum of the sampled signals. This produces a clear peak at the fundamental frequency, with the amplitude of the signals of interest. The values are divided and multiplied by the feedback resistance to obtain the impedance value. The block diagram of the LabVIEW application used to perform this computation is presented in Fig. 6.8.

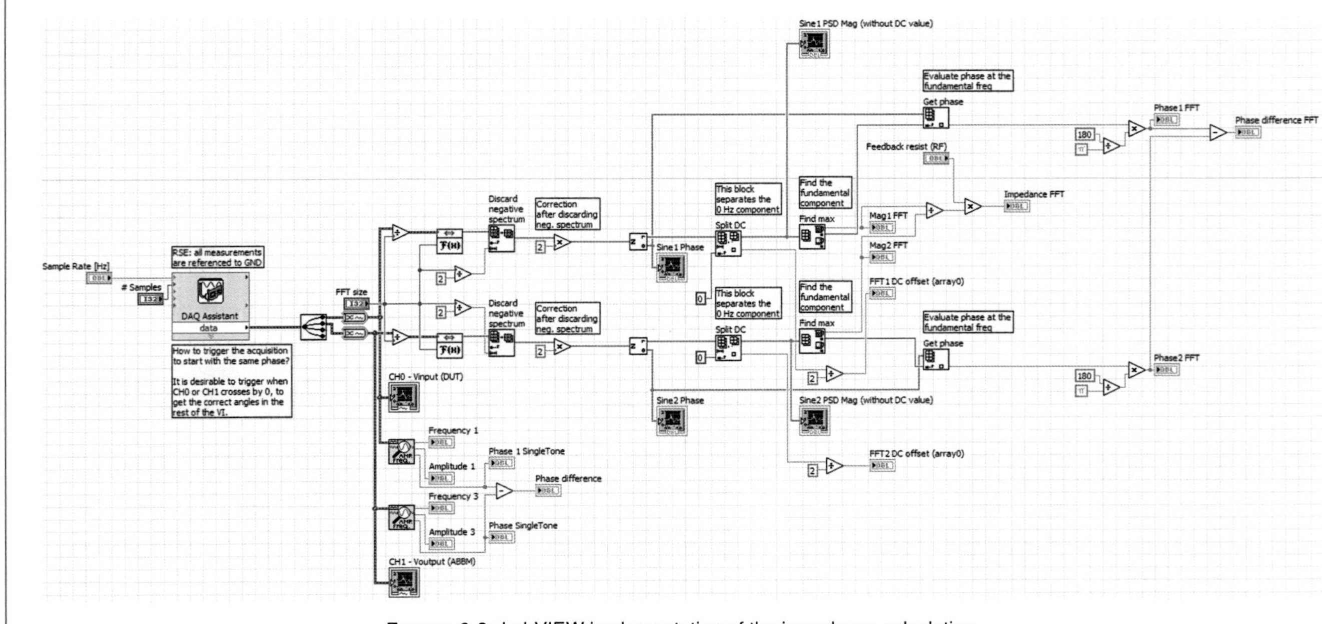

FIGURE 6.8: LabVIEW implementation of the impedance calculation.

6.3.1 Graphical user interface

The user interface of the LabVIEW application is shown in Fig. 6.9. This virtual instrument measures the input and output signal parameters (frequency, magnitude, phase and DC offset) and enables instantaneous calculation of the DUT impedance, in magnitude and phase.

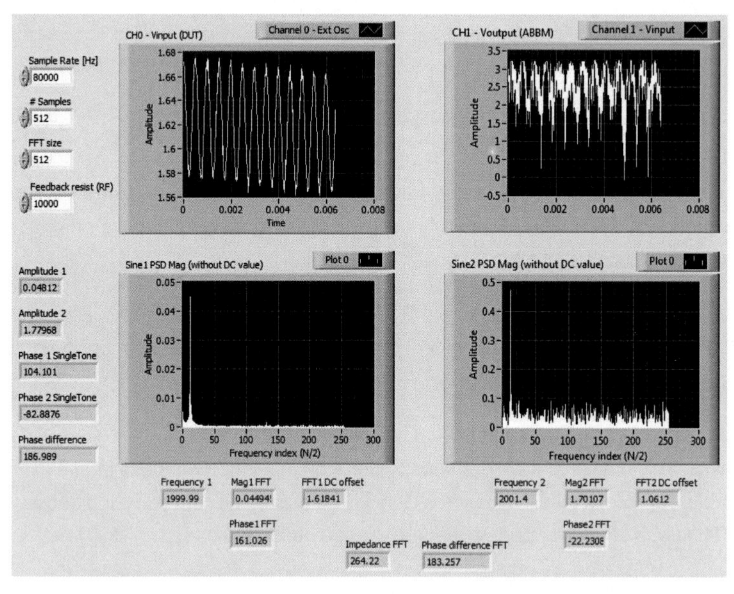

FIGURE 6.9: Graphical user interface for the impedance measurements.

The plots at the upper part are the Vinput and Voutput signals of the ASIC, connected at the channels 1 and 2 of the DAQ interface, respectively. The bottom graphs are the Power Spectral Density (PSD) magnitudes. Horizontal scale ranges from 0 to N/2 according to the DFT theory, using a FFT size of N = 512 samples.

The user must select the sampling rate of the analog channels, the number of samples to obtain from the acquisition board, and the FFT size which generally is equal to the number of samples taken. The user must also enter the value of the feedback resistance programmed in the ASIC.

The main results of the analysis are the impedance magnitude (labeled as Impedance FFT) and phase (labeled as Phase difference FFT), shown at the bottom section of Fig. 6.9. These values are stored periodically in a text file.

6.3.2 Discrete Fourier Transform

The Discrete Fourier Transform is applied to convert the time-domain signals to the frequency-domain, following the analysis equations:

$$\tilde{X}[k] = \sum_{n=0}^{N-1} \tilde{x}[n]e^{-j2\pi kn/N} \tag{6.3}$$

$$\tilde{X}[k] = \sum_{n=0}^{N-1} \tilde{x}[n]\cos(2\pi ki/N) + \sum_{n=0}^{N-1} \tilde{x}[n]\sin(2\pi ki/N) \tag{6.4}$$

The sampled data x(n) consists of two arrays of N=512 values, with a sampling frequency of Fs=80 kHz. One array contains the Vinput signal and the other holds the Voutput signal in the time domain.

The frequency data X(k) has a real ReX[k] and a complex ImX[k] part, and on this VI they are converted to the polar counterparts and represented as magnitude X[k] and phase θ[k].

Each part of the frequency spectrum has N/2 + 1 values, for a total set of N + 2 data points. This means that the frequency index k on the graphics ranges from 0 to N/2. In this example, it takes values from 0 to 256. For the selected sampling frequency of 80 kHz, it corresponds to a frequency range from 0 to 40 kHz.

The *DC offset* of each of the sine waves is stored in the frequency index k=0, and it is then removed from the plots, to be able to analyze the rest of the power spectrum.

The *magnitude* of the fundamental component of the signal is the largest peak in the frequency domain plot, after removing the DC component. The power spectrum is analyzed to find the maximum value, and the index k is stored. This spectrum also shows harmonics which are discarded.

The *phase* of the signal is obtained by evaluating the phase spectrum at the fundamental frequency, given by the index k. The rest of the phase spectrum outside this index is not valid, and it is composed of random noise with values between $-\pi$ and π.

6.4 Impedance measurements

In this experiment, the ASIC is used to measure constant impedances in the four different sensitivity ranges. The ASIC has four different scales (Rf = 10 kΩ, 100 kΩ, 1 MΩ and 10 MΩ) which need to be selected depending on the connected DUT, as described in the operation ranges from Table 6.3.

For the first two scales, an input voltage of 100 mV was selected, since this voltage would result in a feedback current in the range of 1 µA. For a higher impedance range of 1 MΩ, this current becomes smaller and the output voltage falls into the noise level. The input voltage was increased to 2.0 Vpp for the measurements in the 1 MΩ range, and to 3.0 Vpp for the 10 MΩ range, which is close to the supply voltage of the chip. This limited the maximum measurable impedance to 28 MΩ.

The ASIC produces an output voltage for each of the measured impedances. The results of the measurement are presented in Fig. 6.10. The blue lines marked with asterisk are the measurements, and the orange solid lines are the curves obtained by calculating the output voltage with known R_f, R_{DUT} and input voltage.

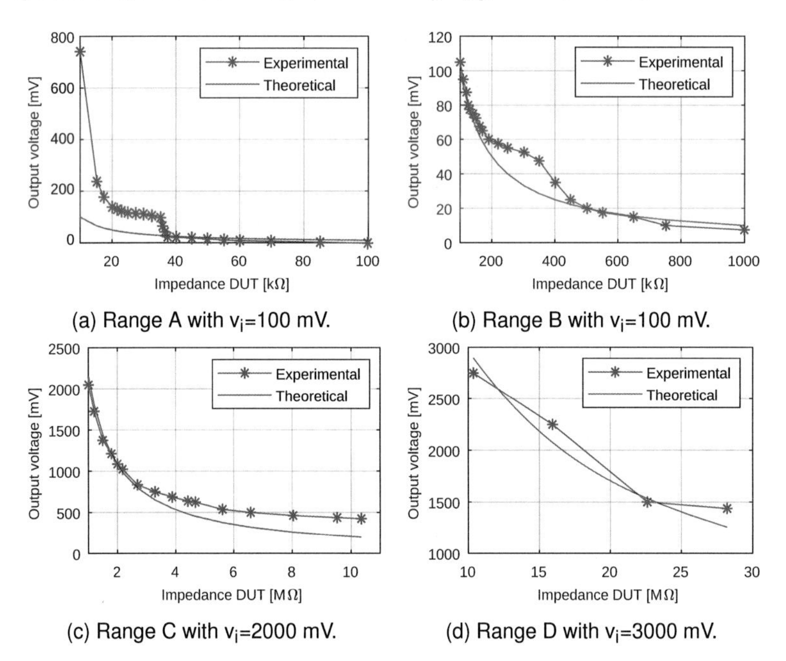

(a) Range A with v_i=100 mV.

(b) Range B with v_i=100 mV.

(c) Range C with v_i=2000 mV.

(d) Range D with v_i=3000 mV.

FIGURE 6.10: Measurement of resistances in the four operation ranges.

6.5 Phase measurements

This section measures the phase of known RC series elements, by keeping a constant resistance and varying the capacitance. This sweeps the phase angle from 0 up to almost 80 degrees. Results are plotted in Fig. 6.11. The plot shows the measured phase as a function of the imaginary part of the capacitance.

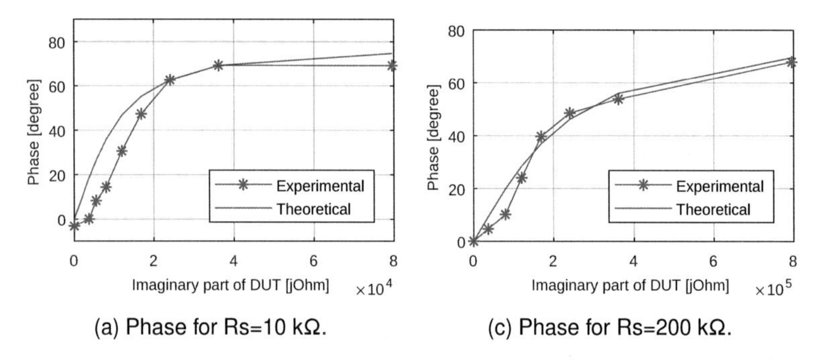

(a) Phase for Rs=10 kΩ. (c) Phase for Rs=200 kΩ.

FIGURE 6.11: Measurement of phase angles in the range B and range C, obtained by keeping the series resistance constant while varying the capacitance.

The measurements from output voltage in Fig. 6.10 and from the phase in Fig. 6.11 show that the impedance can be determined with reasonable accuracy, since the measured quantities are close to the expected values.

The curves are taken under the DC operating point conditions shown in Table 6.4.

TABLE 6.4: DC operating point of the ASIC for the four ranges of operation.

Range	Input voltage	V_{DD}	I_{DD}	I_{DUT}
A	100 mV	3.3 V	6.76 mA	10 µA
B	100 mV	3.3 V	5.56 mA	1 µA
C	2 V	3.3 V	3.24 mA	2 µA
D	3 V	3.3 V	3.24 mA	0.3 µA

This shows that the portable device consumes power in the range from 10.69 mW to 22.31 mW which is a relatively good power consumption.

In the following sections, the impedance system is used to perform biological experiments with yeast cells and porcine chondrocytes.

6.6 Biological experiments

The ASIC was programmed and configured as follows:

- An external oscillator was used in order to have a better tunning and also to control the DC offset voltage Voff.
- The stimulation voltage is of 100 mV peak to peak.
- The stimulation frequency is of 2 kHz.
- The feedback resistance is of 10 kΩ since the impedance is expected in the range from 10 kΩ to 100 kΩ.
- A series resistor of 20 kΩ was used in all the measurements, since the ASIC cannot detect impedances of less than 10 kΩ without entering a saturation condition.

6.6.1 Experiments with yeast

The ASIC is used to monitor the impedance of yeast cultures growing in a well from a 384-well plate, using two pinheads as electrodes. The impedance curves for eight samples are shown in Fig. 6.12, taking one data point per second.

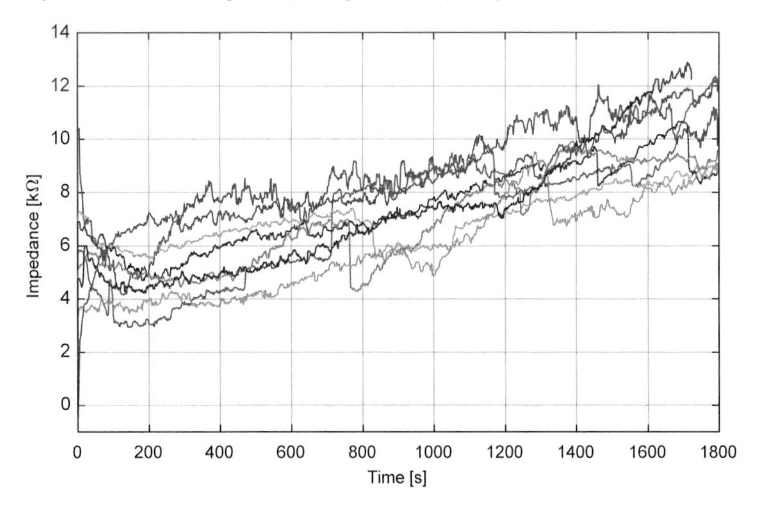

FIGURE 6.12: Impedance of eight different yeast cell cultures, measured with the portable system. Each curve describes an independent experiment.

The experiment is repeated using sugar in the cell suspension. The impedance of water, yeast and yeast with sugar is shown in Fig. 6.13. Yeast cultures with sugar grow faster and exhibit a higher impedance, as shown in Fig. 6.14.

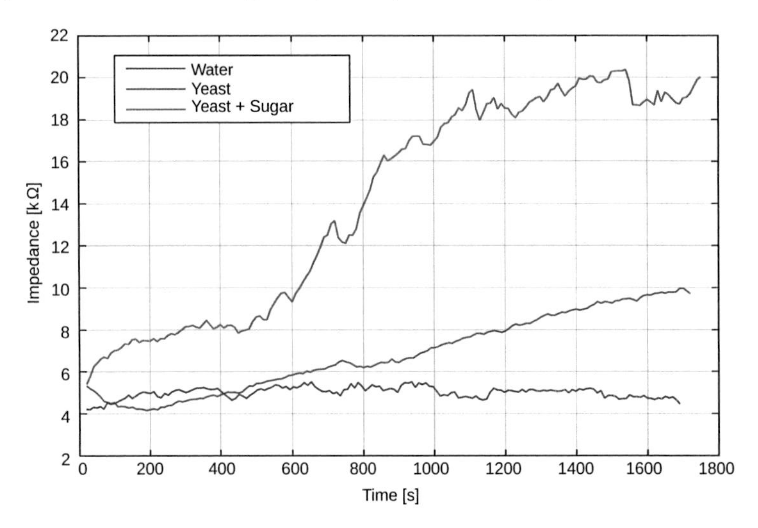

FIGURE 6.13: Impedance of water (blue), yeast wth sugar (green) and yeast without sugar (red).

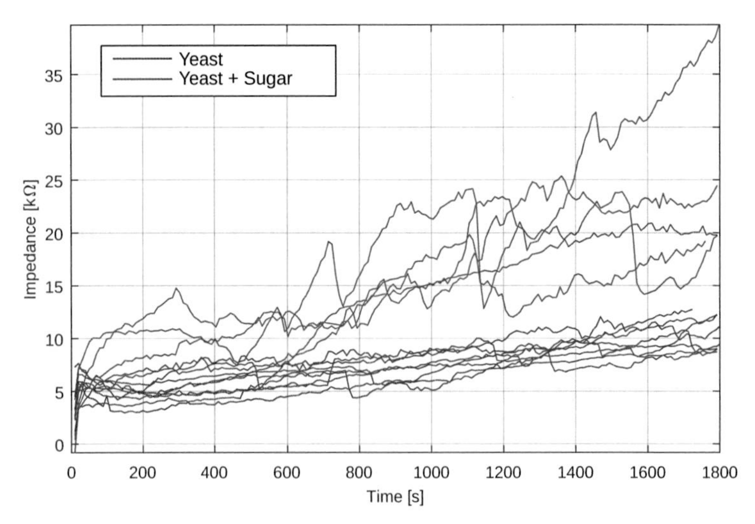

FIGURE 6.14: Impedance of yeast cells with sugar (red) and without sugar (blue) [46].

6.6.2 Experiments with porcine chondrocytes

The ASIC is also used to monitor the impedance of porcine chondrocytes in Dulbecco's Modified Eagle Medium (DMEM). The cells are provided by the Institut für Bioprozess- und Biosystemtechnik.

The initial concentration of cells in the sample is of approximately 1.45×10^6 cells/ml. The cells are placed in a well from a standard 384-well plate, filling a well size of 80 µL. Afterwards, a metallic pinhead connector is immersed in the liquid, in the same way as for the yeast experiments of the previous section.

The impedance of the sample was recorded for more than one hour using the same measurement setup and applied voltages as for the yeast cell experiments. The results are shown in Fig. 6.15.

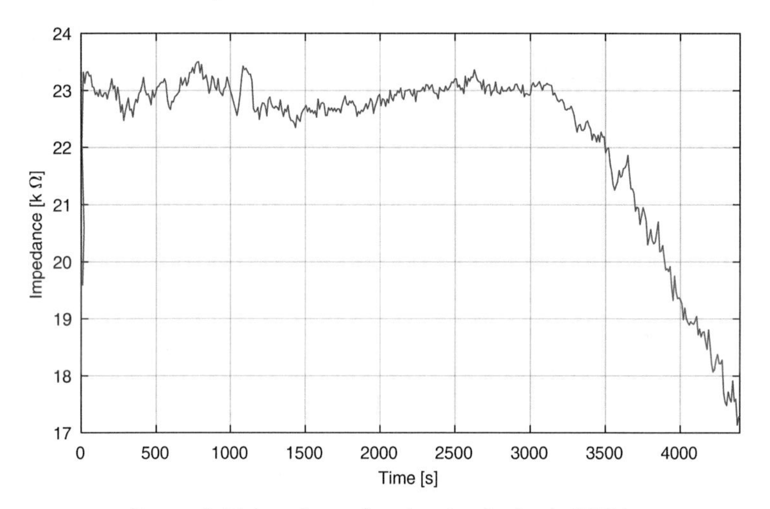

FIGURE 6.15: Impedance of porcine chondrocites in DMEM.

The impedance remains relatively constant for the first 3000 seconds, and then it decreases rapidly. This effect could be related to cell death due to either copper contamination or the sample degradation, since this measurement was not taken under controlled pressure or temperature conditions. The test electrodes used for this measurement were made of copper, and it is possible that copper ions are being released into the medium. Since copper poisoning occurs only above a certain concentration level, this could explain why the impedance was relatively constant during the first 3000 seconds.

6.7 Chapter conclusion

In this chapter, a small mobile system has been realized with an ASIC, reducing the size of the impedance platform further to the centimeter range: the packaged integrated circuit has a PLCC-68 encapsulation of only 16x16 mm. The device has a very small power consumption in the order of 10 to 22 mW which is beneficial for prolonged use in battery-powered environments.

The designed portable system was used to measure impedances in the range from 10 kΩ up to 28 MΩ, providing results for both magnitude and phase with relatively good accuracy. This is achieved by computing the FFT using LabVIEW™ on a personal computer. For a fully mobile system, data could be sent to a smartphone or tablet by the addition of a Bluetooth® module, where the impedance computation can take place given the high-end specifications of recent mobile smartphones.

The system was used to perform cell characterization of yeast. Experiments showed clear differences between cultures with and without sugar in the solution. Addition of sugar accelerated the growth rate, and this was observed in the measurements as a rapid and sustained increase of the impedance.

The ASIC allowed to record also the impedance of a culture of porcine chondrocytes in DMEM. The device detected a rapid decrease of the impedance after 3000 seconds of sampling, which could be explained by the cell deposition at the bottom of the chamber, or by cell death caused by many possible factors such as sample degradation due to bad environmental conditions such as pressure, atmosphere and temperature, or contamination by copper released from the electrodes.

To discard adverse effects, it is recommended to use platinum or Ag/AgCl electrodes in future cell experiments, since the copper pins used on this study may react with the electrolyte and damage the cells. In addition, controlling the temperature and atmosphere usually prolongs cell life.

With the experiments performed in this chapter we conclude that the ASIC can be used to monitor the impedance of biological samples. For most biological applications, the frequency range goes up to 1 MHz, although the H35B4 technology can theoretically support frequencies in several hundreds of MHz. A redesign of the integrated circuit would allow exposing cell cultures to an extended range of frequencies for a broader spectrum of applications.

Chapter 7

High frequency stimulation

The ASIC 2 was designed by P. Vega-Castillo and fabricated in IHP SG13s 130 nm technology. This ASIC measures the magnitude and phase of impedances at frequencies from 30-50 GHz. The top level of this chip is shown in Fig. 7.1. The silicon die includes an isolated test structure for the quadrature oscillator (bottom left), the impedance measurement system (top right) and two extra inductors (top left and bottom right).

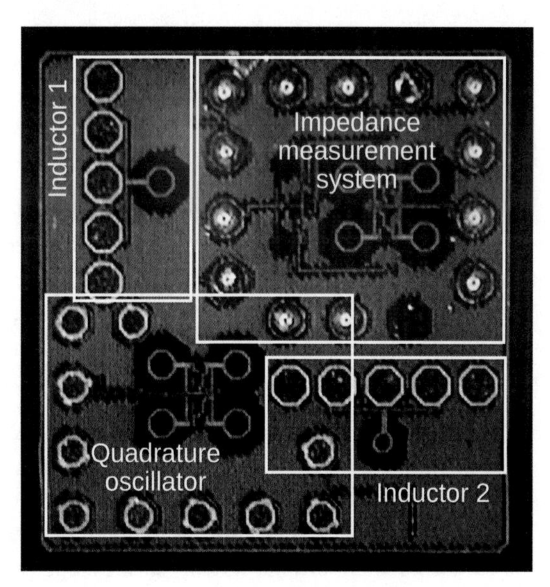

FIGURE 7.1: Photography of the ASIC 2.

7.1 Chip architecture

The ASIC 2 is based on the measurement technique known as frequency response analysis (FRA). It includes a cross-coupled quadrature oscillator, a Cherry-Hooper amplifier, two Gilbert mixers and two RC low-pass filters as shown in Figure 7.2.

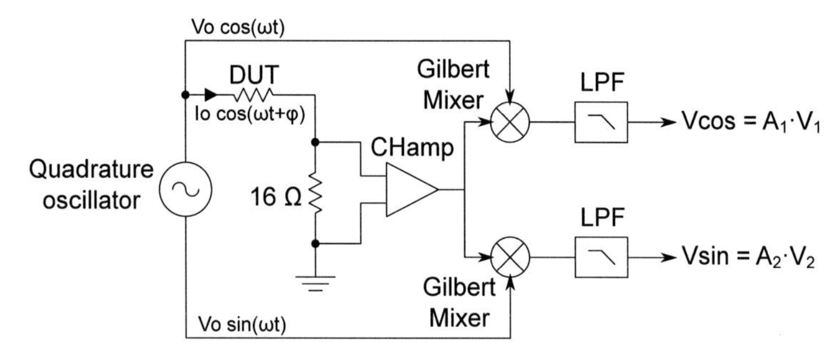

FIGURE 7.2: Topology of the ASIC 2 based on the FRA method.

The outputs *vcos* and *vsin* are DC signals proportional to the real and complex impedance of the device-under-test (DUT).

7.1.1 Quadrature oscillator

The quadrature oscillator is based on the differential LC oscillator topology. An initial design by J. Prada-Rojas was developed using CMOS technology [113]. The specifications of this oscillator are: an output voltage of 100 mVpp, an output coupling of 50 Ω and a frequency range from 1 to 5 GHz. This frequency range was later extended to 30-50 GHz in the final ASIC design by P. Vega-Castillo, by using a silicon-germanium BiCMOS technology.

The characteristic of a quadrature oscillator is the phase difference of 90 degrees between each one of the four outputs. Quadrature in this oscillator is achieved by means of two identical oscillators, interconnected by the output terminals ZeroL, ZeroR, 90L and 90L to the collectors of the transistors Q7, Q8, Q9 and Q10. This connection causes both oscillators to oscillate with a phase shift of exactly 90 degrees between each output signal.

The schematic of the oscillator is shown in Fig. 7.3. The transistors Q11, Q5, Q6, Q12 and Q13 form a current mirror which requires a reference current of 3-4 mA. The actual oscillation is provided by the transistors Q1, Q2, Q3 and Q4.

Considering only one pair of transistors, say Q1 and Q2, there will be one transistor that turns on faster than the other. Assuming Q1 turns on faster, this produces a drain current which charges the varactor until the voltage in the gate of Q2 is enough to activate it. The transistor Q2 turns on, discharging the varactor, which accumulates charge now in the opposite direction, and it causes Q1 to turn off. The process repeats indefinitely, producing the oscillation.

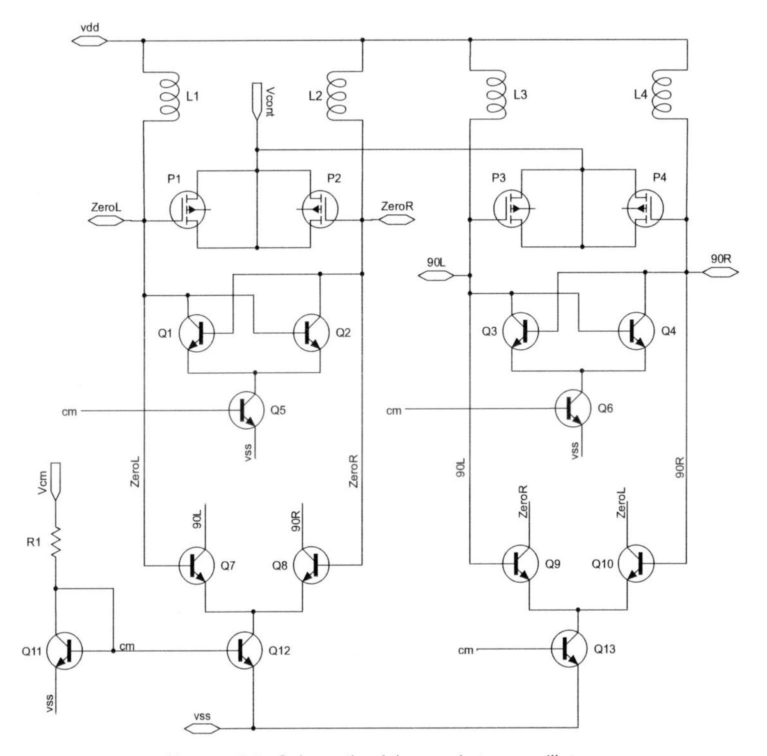

FIGURE 7.3: Schematic of the quadrature oscillator.

The oscillation frequency is set by the values of the inductors L1, L2, L3 and L4, and the capacitance of the varactors P1, P2, P3 and P4. By adjusting the control voltage of the varactors (Vcont), the frequency of this oscillator can be modified.

7.1.2 Emitter followers

Emitter followers are used to provide matching to the circuit. The schematic for all the emitter-followers used in the ASIC 2 is shown in Fig. 7.4.

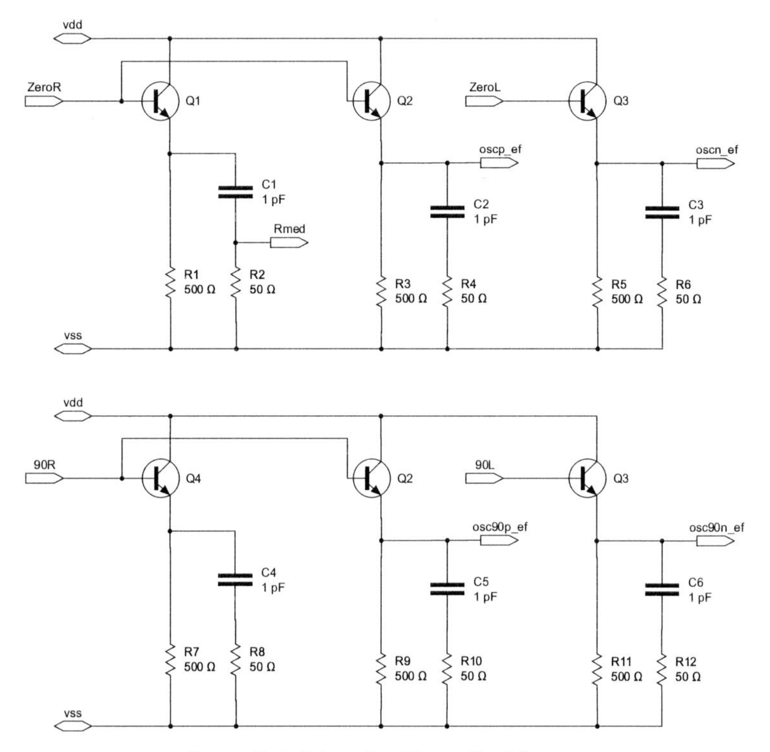

FIGURE 7.4: Schematic of the emitter followers.

The input impedance of an emitter follower is given by $r_{in} = \beta(R_E||R_L)$. According to the schematic, $R_E = 500\Omega$ and $R_L = 50\Omega$. The gain β is given by the physical dimensions of the transistors, which are w=0.12 μm and l=1 μm, as well as the process parameters such as the doping concentrations. The emitter followers are designed to provide matching of 50 Ω to the four outputs of the oscillator.

The output Rmed is connected to the DUT, and the outputs oscp_ef, oscn_ef, osc90p_ef and osc90n_ef are connected to the Gilbert mixers.

7.1.3 Cherry-Hooper amplifier

The schematic of the Cherry-Hooper amplifier is given in Fig. 7.5. The amplifier is used to provide gain to the signal measured by the 16 Ω resistor.

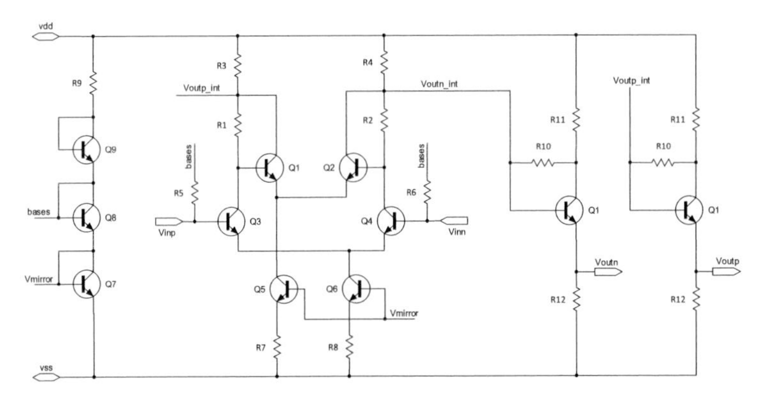

FIGURE 7.5: Schematic of the amplifier.

7.1.4 Gilbert mixers

The Gilbert mixers are shown in Fig. 7.6. These circuits perform multiplication of analog waveforms, and include basic RC low-pass filters at the DC outputs.

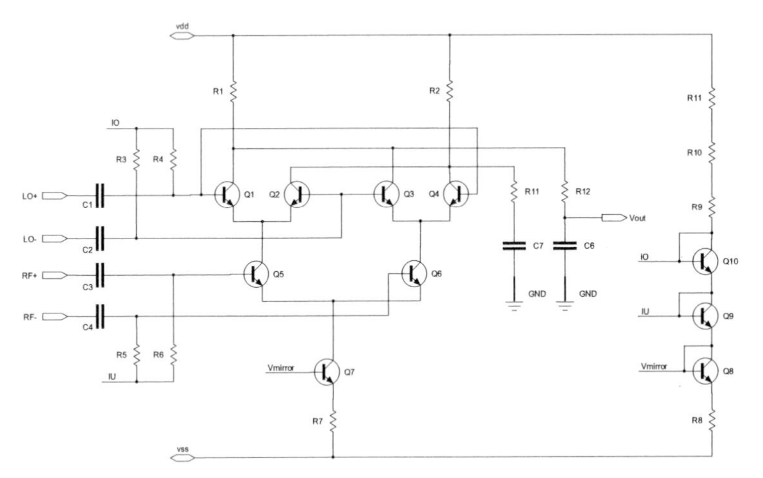

FIGURE 7.6: Schematic of the Gilbert mixer.

7.2 Impedance calculation

The output of the circuit consists of two DC signals, proportional to the cosine and the sine of the current flowing through the DUT.

The output signals are calculated by applying trigonometric identities, as follows:

$$2\cos(\alpha)\cos(\beta) = \cos(\alpha-\beta) + \cos(\alpha+\beta) \tag{7.1}$$
$$2\sin(\alpha)\cos(\beta) = \sin(\alpha+\beta) + \sin(\alpha-\beta) \tag{7.2}$$

$$\begin{aligned} \mathrm{Vcos} &= \mathrm{I_0}\sin(\omega t) \times \sin(\omega t + \phi) && (7.3)\\ &= \frac{\mathrm{I_0}}{2}[\cos(\phi) - \cos(2\omega t + \phi)] && (7.4)\\ &\approx \frac{\mathrm{I_0}}{2}\cos(\phi) && (7.5)\\ &= A_1 v_1 && (7.6) \end{aligned}$$

$$\begin{aligned} \mathrm{Vsin} &= \mathrm{I_0}\cos(\omega t) \times \sin(\omega t + \phi) && (7.7)\\ &= \frac{\mathrm{I_0}}{2}[\sin(\phi) + \sin(2\omega t + \phi)] && (7.8)\\ &\approx \frac{\mathrm{I_0}}{2}\sin(\phi) && (7.9)\\ &= A_2 v_2 && (7.10) \end{aligned}$$

The terms involving $2\omega t$ in eq. (7.4) and eq. (7.8) are filtered by the low-pass filters.

The current I_0 flowing through the DUT is calculated as:

$$I_0 = 2\sqrt{A_1{}^2 v_1{}^2 + A_2{}^2 v_2{}^2} \tag{7.11}$$

The measurement is accurate if the DUT is large compared to the internal 16 Ω resistor. Assuming Z >> 16Ω:

$$|Z| = \frac{V_0}{I_0} = \frac{V_0}{2\sqrt{A_1{}^2 v_1{}^2 + A_2{}^2 v_2{}^2}} \tag{7.12}$$

where V_0 is the oscillator voltage.

The phase angle ϕ is obtained as:

$$\phi = \arccos\left(\frac{2A_1 v_1}{I_0}\right) \tag{7.13}$$

The constants A_1 and A_2 are obtained experimentally by connecting a known calibration load (e.g. R = 50 Ω). The output voltages are measured, and the current I_0 and phase ϕ are known for this load.

7.3 Simulations of an RC impedance load

To demonstrate the process of impedance calculation, consider the following simulation of the ASIC with a parallel RC load. The oscillation frequency is of 41.926 GHz, using V_{cont}=-0.45 V and I_{CM}=3 mA. The results of the simulation are observed in Fig. 7.7, showing the oscillator voltage and the current flowing through the DUT.

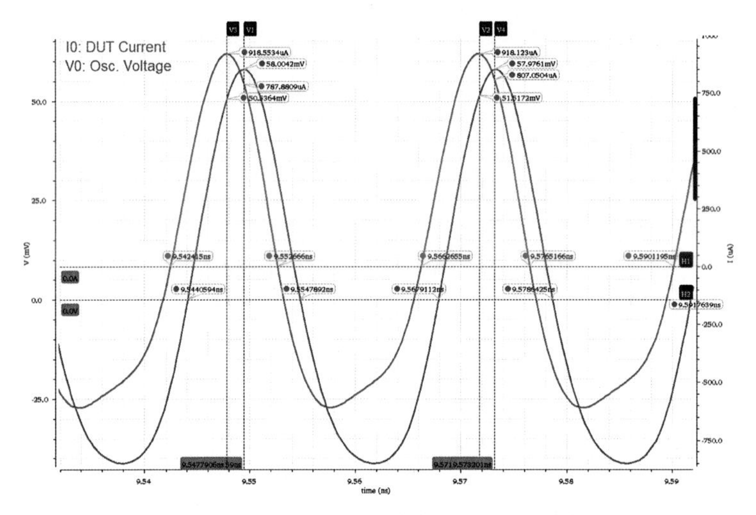

FIGURE 7.7: Simulation of the ASIC with a parallel RC load.

The characteristics of the parallel RC load are described in Table 7.1, and the values read from the simulation are summarized in Table 7.2.

TABLE 7.1: Specifications of the parallel RC load.

Parameter	Value
R	70.7 Ω
XC	-j69.47 Ω
Z_{DUT}	$49.55\ \Omega\angle -45.5°$
Z_{DUT}+16 Ω	$61.82\ \Omega\angle -34.86°$

TABLE 7.2: Values read from the simulation.

Parameter	Value
V0	58.00 mV
I0	918.55 uA
T	23.8533 ps
ΔT	2.1259 ps
ϕ	32.0846°
v1	254.792 mV
v2	307.937 mV

The gains A_1 and A_2 are theoretically calculated from the measurement as:

$$\begin{aligned} A_1 v_1 &= \frac{I_0}{2}\cos\phi \Rightarrow A_1 = \frac{I_0}{2v_1}\cos\phi \\ A_1 &= \frac{918.55\mu A}{2\cdot 254.79 mV}\cos(-32.08°) \\ A_1 &= 1.527\times 10^{-3} A/V \end{aligned}$$

$$\begin{aligned} A_2 v_2 &= \frac{I_0}{2}\sin\phi \Rightarrow A_2 = \frac{I_0}{2v_2}\sin\phi \\ A_2 &= \frac{918.55\mu A}{2\cdot 307.94 mV}\cos(-32.08°) \\ A_2 &= 792.11\times 10^{-6} A/V \end{aligned}$$

These constants are needed to calculate the impedance of the DUT in the experiments with the ASIC, since there is no way to measure the amplitude of I_0 and V_0 in the experimental setup. Assuming A_1 and A_2 are already available, the following impedance calculation is performed:

$$|Z| = \frac{V_0}{I_0} = \frac{V_0}{2\sqrt{A_1{}^2 v_1{}^2 + A_2{}^2 v_2{}^2}} = \frac{58.00mV}{918.55\mu A} = 63.14\Omega \tag{7.14}$$

The phase is calculated as:

$$\phi = \arccos\left(\frac{2A_1 v_1}{I_0}\right) = \arccos\left(\frac{2\cdot(1.527\times 10^{-3})(254.79mV)}{918.55\mu A}\right) = -32.09° \tag{7.15}$$

The calculations given in this section show that determination of magnitude and phase can be done within good reasonable accuracy for RC loads.

7.4 DC operating point of the VCO

Measurements of the DC operating point were carried out using a probe station with Karl Suss PH100 needles, and the Keithley SCS4200 SMU. The experimental setup is shown in Fig. 7.8.

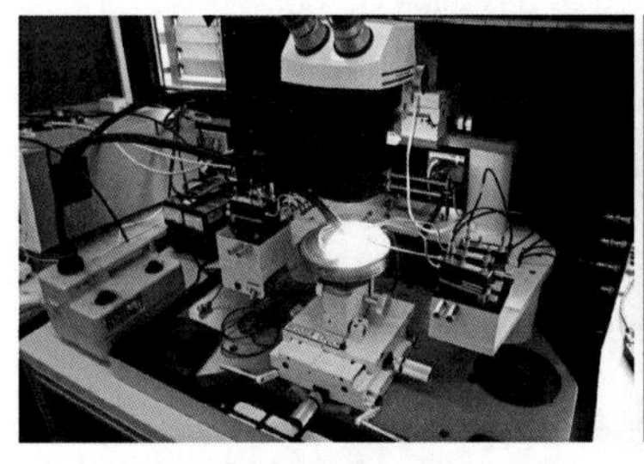
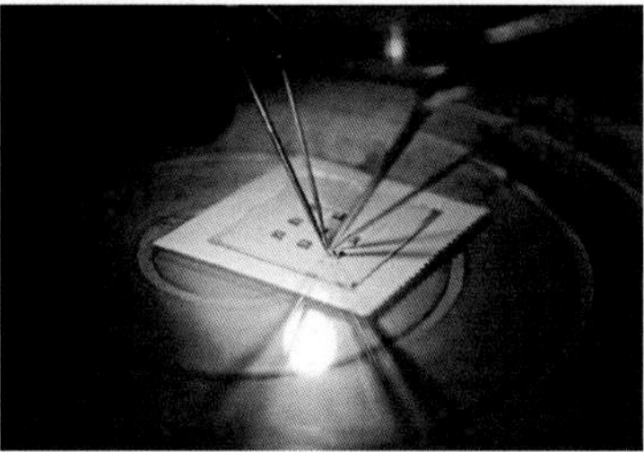

FIGURE 7.8: Measurement of the DC operating point of the VCO.

Operating point measurements of the quadrature oscillator were carried out using the oscillator from the full impedance measurement system, instead of the isolated VCO, to estimate the total power consumption of the device. The operating point for successful oscillation is shown in Table 7.3.

TABLE 7.3: DC operating point for the ASIC 2.

Parameter	Theoretical	Measured
VDD	-1.65V	-1.65V
VSS	-3.3V	-3.3V
VCONT	-0.45V to -1.65V	-0.55V to -1.65V
VCM	+0.6V	+1.65V
GND	0	0
IDD	20.98 mA	18.08 mA
ISS	24.02 mA	23.37 mA
ICONT	0	1.42 mA*
ICM	3mA	3.87 mA
IGND	0	0

*Initially ICONT=260 µA, increasing with the total time of chip operation.

The chip consumes a total of 23.37 mA using -3.3 V which translates into a power of 77.12 mW. The current is consistent with the KCL: $I_{SS} = I_{DD} + I_{CONT} + I_{CM}$.

7.5 Inductor S-parameters

The S-Parameters measure the transmitted and reflected power of a signal when it travels trough a two-port component such as an inductor. This section compares the simulated S-parameters of an inductor with the measurement of the parameters on the test inductors of the ASIC, performed by using the HP 8051 VNA.

7.5.1 Inductor simulation using HFSS

The simulation of the 125 pH inductor is constructed in Ansys HFSS by directly importing the geometry from Cadence. The model is shown in Fig. 7.9.

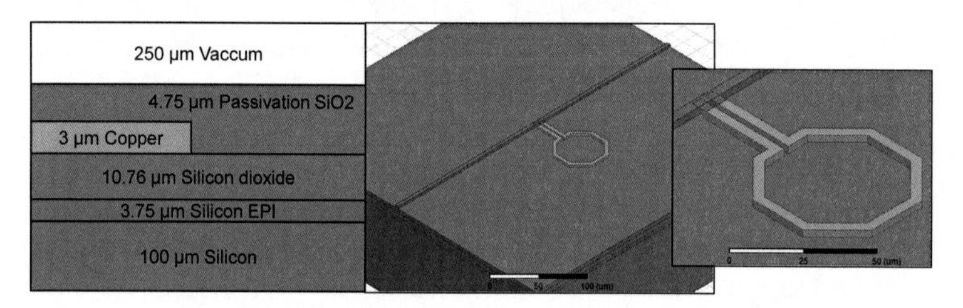

FIGURE 7.9: Model of a 125 pH inductor in Ansys HFSS.

The simulation of this inductor yields the results of Fig. 7.10. The left plot shows the S11 parameter with the reflected power, and the right plot shows the S21 parameter with the transmitted power.

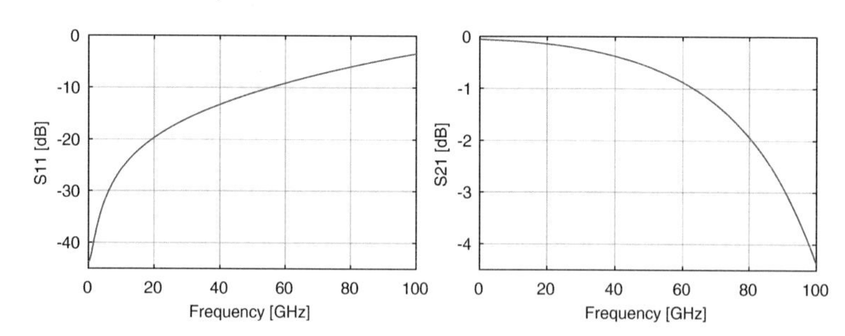

FIGURE 7.10: Simulated S-parameters of the 125 pH inductor.

In the next section, a VNA is used to measure these parameters directly from the test inductors integrated on the ASIC.

7.5.2 SOLT Callibration

Before measuring the S-parameters of any device, the VNA must be calibrated to eliminate the effect of the parasitics from the setup. This is achieved by using a calibration substrate, shown in Fig. 7.11. The SOLT calibration (Short, Open, Load, Thru) is performed, and the results are presented in Fig. 7.12.

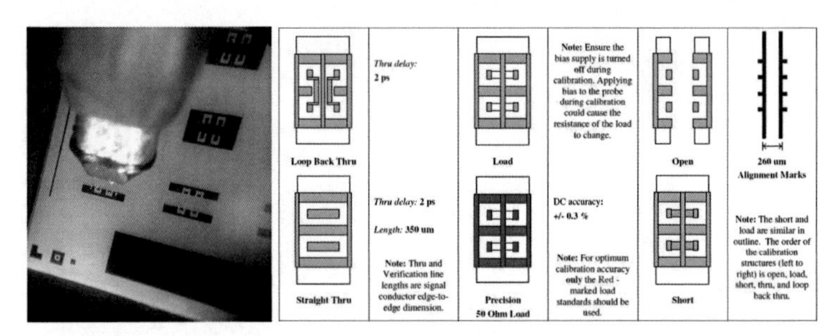

FIGURE 7.11: Calibration substrate specifications, model ISS 129-239 by Cascade Microtech.

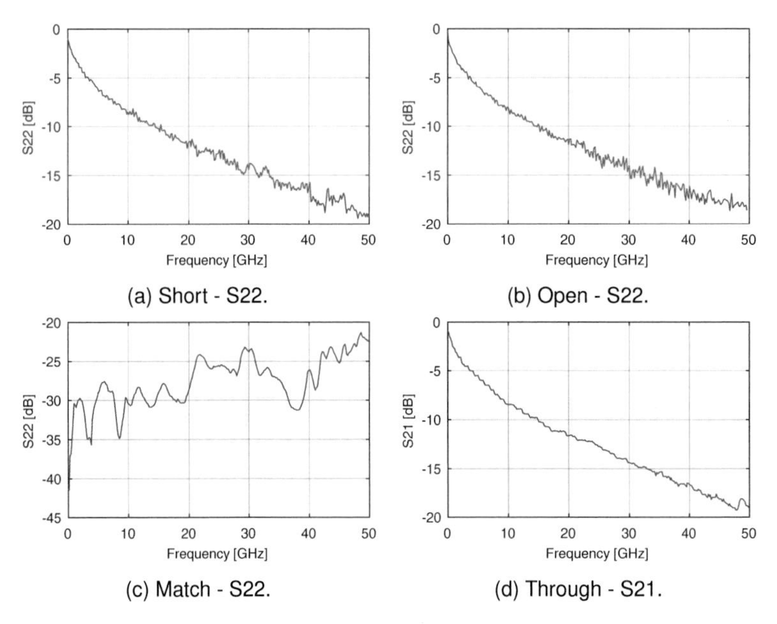

FIGURE 7.12: SOLT calibration results of the VNA 8051.

7.5.3 Inductor measurements

With the VNA already calibrated, the S-parameters of the two on-chip inductors were measured in the range from DC to 50 GHz. The experimental setup is presented in Fig. 7.13, and the results are plotted in Fig. 7.14.

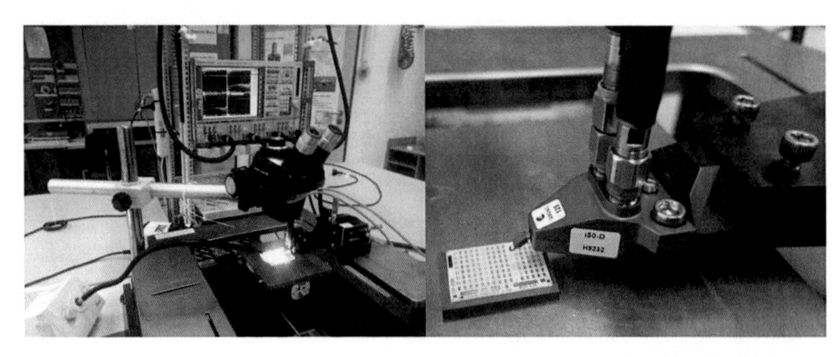

FIGURE 7.13: (a) Vector Network Analyzer and (b) Infinity Probe with the calibration substrate.

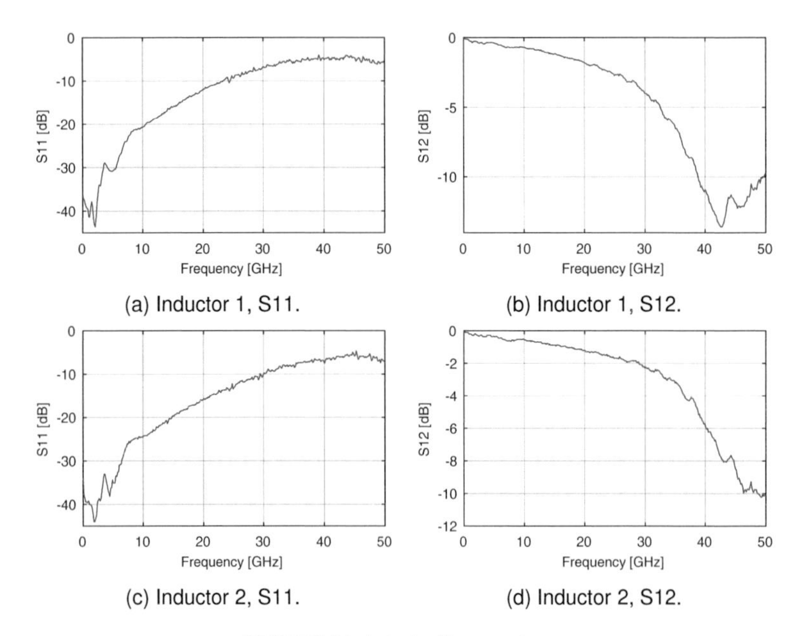

(a) Inductor 1, S11.

(b) Inductor 1, S12.

(c) Inductor 2, S11.

(d) Inductor 2, S12.

FIGURE 7.14: Inductor S-parameters.

7.6 Oscillator measurements

The output voltage of the voltage-controlled oscillator (VCO) is measured with needles through the pads in the silicon die.

Measurements were carried out with an Infinity probe GSGSG-125 connected to a Rhode&Schwarz FSUP50 Signal Source Analyzer. Due to the lack of space in the die, one of the signal pins of the probe was placed on top of the LOAD+ pad, while the other signal pin is left floating. In addition to the HF probe, four extra needles are required to provide VSS, VDD, ICM and VCONT to the oscillator.

The distance between adjacent pads is 150 µm for this measurement. For easier manipulation of the needles, a ceramic PCB was designed and fabricated at the Forschungswerkstatt Elektrotechnik (FWE) of the TUHH with the assistance of colleagues from the Institute of High Frequency. The PCB was fabricated in a flexible ceramic substrate with a single layer of gold coating.

Nine silicon dies were fixed to this ceramic PCB with an epoxy compound to ensure mechanical stability. Silicon dies were glued to the ceramic PCB using epoxy compound from Epoxy Technology (EPO-TEK H37-MP) and baked in the oven at 150 °C for 30 minutes.

The PCB and needle connections for this measurement is shown in Fig. 7.15. This measurement is performed in one of the two VCOs present in the die; specifically, on the VCO which is part of the impedance measurement system.

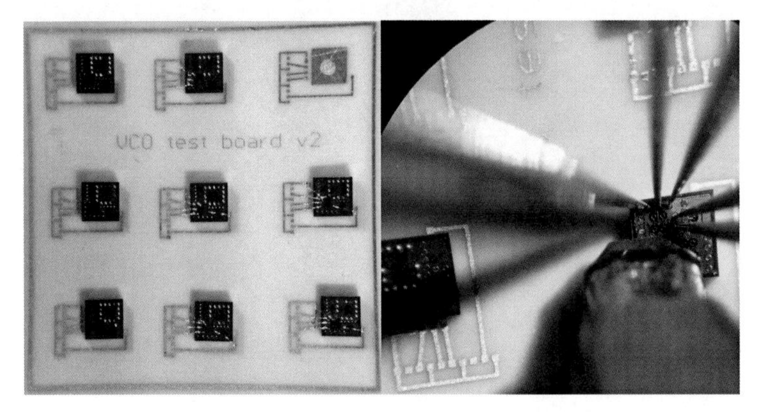

FIGURE 7.15: Test board and experimental setup for the VCO oscillator output measurement, showing four DC needles and a HF Infinity Probe.

The designed PCB also includes connectors for attaching a 10-contact DC probe from Cascade Microtech. This DC probe would avoid the use of needles for the low-frequency terminals of the standalone oscillator (VSS, VDD, VCONT, VCM, GND). Wire bonding was carried out at the High Frequency Institute, using standard gold wires. The final result is shown in Fig. 7.16.

FIGURE 7.16: Wire bonding of DC connections to the isolated VCO.

Additional pictures of the experiment are found in Fig. 7.17.

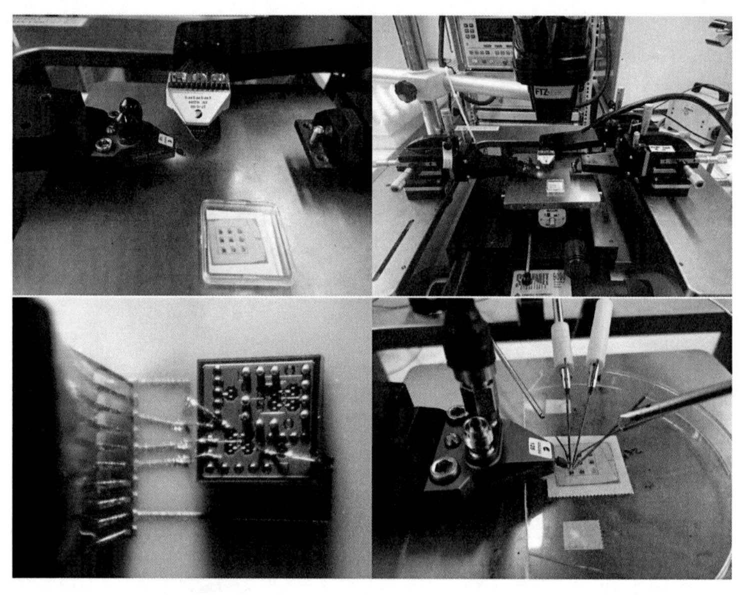

FIGURE 7.17: Additional pictures of the experiments, showing the probing station, the HF probe and the DC probe.

The results of the oscillator measurement are presented in Fig. 7.18. In this particular figure, the output frequency is of 39.37 GHz, with an amplitude of -46.10 dBm, which corresponds to a peak amplitude of 1.567 mV.

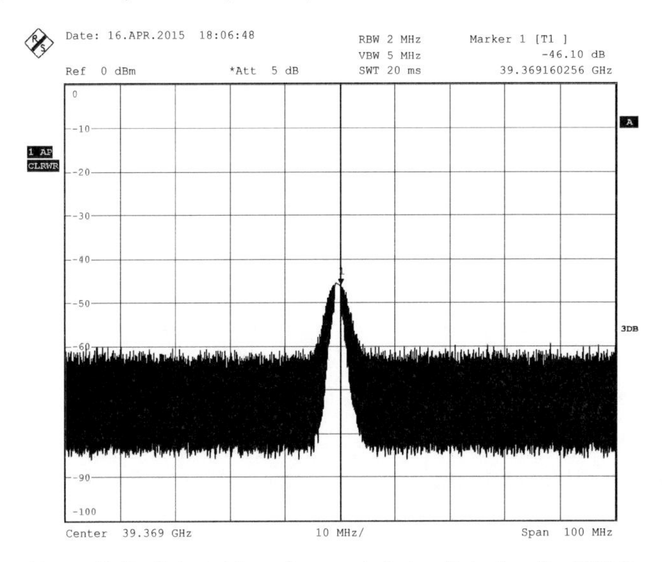

FIGURE 7.18: Output of the voltage-controlled oscillator from the ASIC 2.

Variation of parameters such as the reference current I_{CM} or the control voltage of the varactors V_{CONT} gives some option to tune the performance of the oscillator, specifically the amplitude and the oscillation frequency.

The amplitude of the oscillation varies from -52 to -45 dBm by adjusting the current on the V_{CM} terminal in the range from 3 to 4 mA. This corresponds to a peak amplitude ranging from 561.68 µV up to 1.78 mV. This is far from the designed 100 mV, or the expected 58 mV from the simulation shown earlier in this chapter. However, it is not unexpected, since the connection of the HF probe, wiring and external equipment introduces additional mismatches that cannot be easily compensated on the chip side.

The oscillation frequency varies from 38.94 to 39.65 GHz when the control voltage is adjusted from -0.45 V to -1.65 V respectively. This is also far from the design specifications, where the oscillation frequency could be adjusted from 30 to 50 GHz, and the possible reason is that the pad parasitics are much higher than the expected capacitance of the varactor.

7.7 Chapter conclusion

In this chapter we tested the main components of a high-frequency impedance measurement ASIC, specifically the DC operating point, the S-parameters of the integrated inductors, and the oscillator frequency and output voltage of the quadrature oscillator.

The DC operating point was measured and shown in Table 7.3. The voltages and current consumption of the ASIC are within the expected values. Furthermore, it is observed that the chip consumes a total power of 3.3 V×23.37 mA=77.12 mW. This considers both the oscillator and the impedance measurement circuitry. Compared with the ASIC 1, the power consumption increased almost four times, but it is expected since this chip operates in a much higher frequency range. Still, the measured power consumption is more than acceptable for most portable applications.

The S-parameters of the inductors were simulated and also measured directly by using a VNA. The results are as expected, showing almost the same frequency response for the S11 parameter in both simulation and measurement. The S21 parameter is less in practice, since transmission in the physical sensor was expected to be lower, but still shows a transmitted power above -10 dBm for frequencies up to 40 GHz which is a positive result.

The quadrature voltage controlled oscillator (VCO) is working and the oscillation was measured by a source analyzer at a frequency of 39.37 GHz. This frequency can be tuned from 38.94 to 39.65 GHz by using the control voltage of the varactors, which modifies their capacitance and therefore the resonant frequency of the LC elements. Although some modification of the frequency was possible, the full range from 30 GHz to 50 GHz was not achieved. This could be due to parasitic elements in the control pad of the varactor, which may dominate over the varactor capacitance.

The output voltage of the oscillator was found to be low, ranging from 561.68 µV up to 1.78 mV when tuned by the reference current of the current mirrors. The expected oscillation voltage was of 100 mV by design and 58 mV after simulation. These amplitudes were not achieved. There are two possible reasons for this, explained as follows.

The first possible reason is the unbalancing of the oscillator during the measurements with the Infinity Probe. This probe was used to contact only one of the two outputs of the oscillator, leaving the other output floating. An optimal setup would require placing a termination of 50 Ω in one side of the probe, and connecting the other side to the VNA which provides proper matching. This was impossible to achieve in the present setup due to space limitations on the chip surface, since there was no possible way to accommodate the two active terminals of the probe and the four needles required for DC power supply.

The second explanation for the low oscillation voltage is due to the matching of the oscillator to the external load and then to the 16 Ω internal resistor. This branch would be balanced only if the DUT is of 34 Ω. Looking at the schematics of the emitter followers (EF) (see Fig. 7.4), it is clear that the four outputs of the VCO have different loads: two are connected to one EF, and two have two EFs. In addition, the capacitance of every pad adds to the mismatch, and must be considered in the redesign by electromagnetic modeling and full wave simulations.

The majority of the issues encountered in the present chapter were addressed by the chip designers during the development of the ASIC 3. Up to this point we have demonstrated that is possible to integrate a full impedance measurement system on an integrated circuit, including a working oscillator in millimeter-wave frequencies, integrating the inductors on the chip which eliminates the need of any external component.

The next iteration of the design is presented in the next chapter.

Chapter 8

High frequency stimulation redesign

A redesign of the ASIC 2 was carried out by P. Budyakov using the same 130 nm IHP SG13S technology. This ASIC was shipped without any packaging or encapsulation. A chip photography is shown in Fig. 8.1.

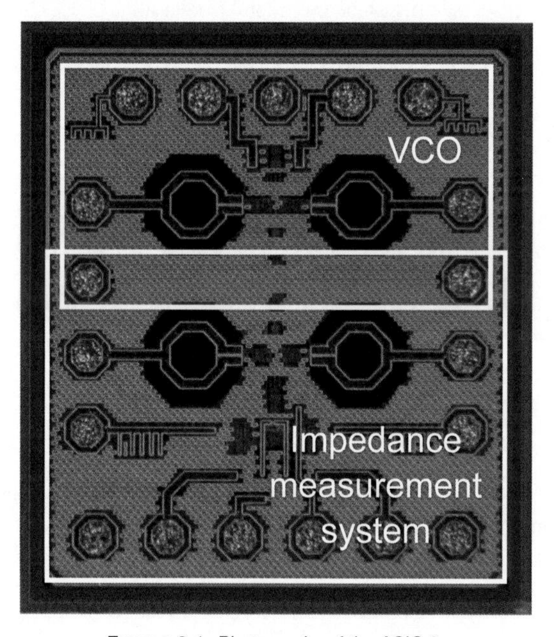

FIGURE 8.1: Photography of the ASIC 3.

8.1 Chip architecture and simulations

The topology of this chip is exactly the same as of the ASIC 2, with some additions to improve the matching and the overall performance of the circuit. The schematics are not included in the dissertation, since the goal of this chapter is to address the assembly of test fixtures for this device. However, a brief description of the most important changes is given as follows.

The oscillator now includes additional capacitances for improved stability, and each output of the VCO is connected to two emitter followers in a symmetric way, reducing the mismatch to the rest of the chip components.

A major improvement over the previous design is the inclusion of full wave electromagnetic simulations on the design process, performed using Keysight® Advanced Design System™. Another important change is the use of only two inductors with a center tap, instead of the four used in ASIC 2.

The Gilbert mixers of this ASIC include two fully differential outputs: Vout1P-Vout1N and Vout2P-VoutN. These signal describe the impedance of the DUT in the same way of Vcos and Vsin for the previous design. By using Equations 7.1 to 7.13, the impedance of the load may be calculated in the same way.

The circuit is simulated to obtain the expected results for the full range of impedance. For this purpose, the resistance of the DUT is swept from 1 mΩ up to 10 MΩ in the simulation, and the output voltages are plotted in Fig. 8.2.

Observe that for loads higher than 10 kΩ, the four output channels have a DC value of approximately 3.0 V which gives a differential output of zero. On the contrary, loads lower than 10 Ω produce saturation of the four channels, which now give the maximum voltage difference.

Based on this simulation, open- and short-circuit loads should give the maximum and minimum output voltage differences across the four channels. The expected results are given in Table 8.1.

TABLE 8.1: Expected results for open- and short-circuit loads.

Test	Vout1P-Vout1N	Vout2P-Vout2N
Short-circuit	320 mV	130 mV
Open-circuit	0 V	0 V

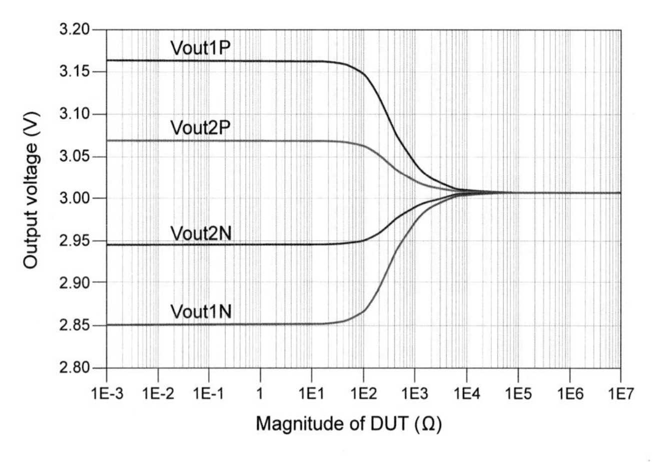

FIGURE 8.2: Output voltage of the ASIC 3 for resistive DUT impedances.

From this figure, the detectable range of impedances is from 10 Ω up to 10 kΩ, since values outside this window produce a saturated output. In other words, impedances below 10 Ω produce no detectable output, and the same occurs if the DUT has an impedance that exceeds 10 kΩ.

8.2 Assembly technologies for bare die ASICs

Since the ASIC 2 and ASIC 3 operate at frequencies higher than 40 GHz, the chips cannot be packaged because of the large parasitics of the wires and encapsulation. Therefore, the assembly requires advanced techniques such as flip-chip and wire bonding.

These are industrial methods, used for large production volumes, and the companies often have test runs for adjusting the parameters such as the bonding temperature curve, or the pressure applied to the parts. In this section we describe both methods, and compare them by simulation using HFSS.

8.2.1 Flip chip bonding

Flip-chip bonding is an assembly technology intended for interconnections between an integrated circuit and another circuitry such as a Printed Circuit Board (PCB). It is based on solder bumps that are previously deposited on the pad of the silicon die. The bumps melt under specific temperature and pressure conditions, and form an electric connection by reflow soldering. It is called 'flip-chip' because the silicon die is placed upside down, with the pads facing the surface of the PCB.

The assembly of a test board for impedance measurement using the ASIC was pursued by using the flip-chip assembly method using the Finetech FINEPLACER Pico ma. The solder bumps on the ASIC are made of an alloy of AuSn 80/20. They are intended to be soldered on gold-coated surfaces. For this purpose, the target PCBs were coated with a thin layer of gold by chemical deposition at the Institute for High Frequency.

After this procedure, bonding is performed according to the steps illustrated in Fig. 8.3. The silicon die is picked up by suction, aligned with the PCB tracks by using a dual camera system, then placed on the PCB site, and soldered by heating the PCB according to a temperature profile.

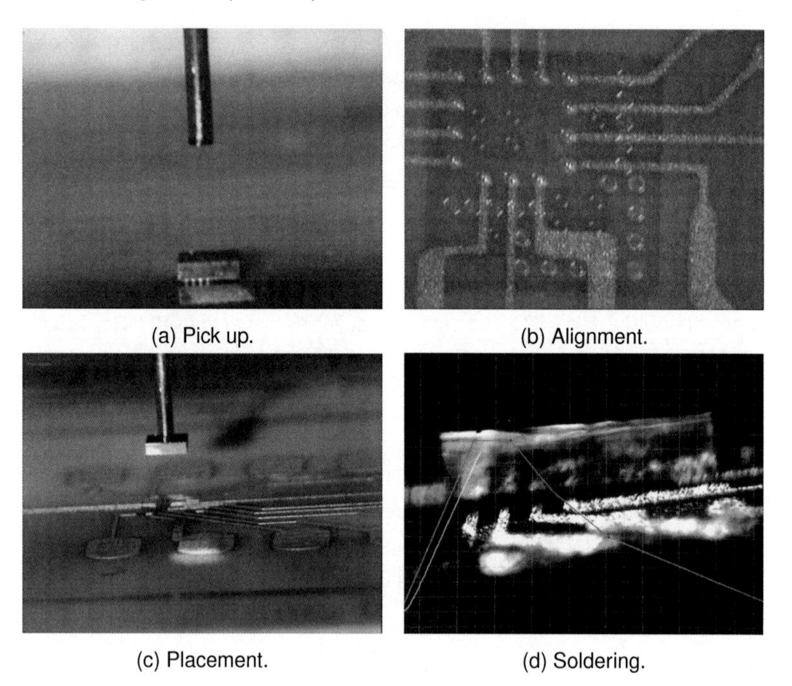

(a) Pick up. (b) Alignment.

(c) Placement. (d) Soldering.

FIGURE 8.3: Pick up, alignment, placement and soldering of a silicon die by using the flip-chip bonding procedure.

The process parameters for this method are the temperature, soldering time and the applied pressure during soldering.

The temperature curve includes typically a pre-heating step of 100 °C for 90 seconds, followed by a specific slope for the heat increase, until the soldering temperature is achieved. This maximum temperature is kept for 10-20 seconds, and then the equipment blows cold air to accelerate the cooling process. The exact temperature curves are suggested in data sheets from the flip-chip vendors, but each application is different and requires destructive testing to find the optimal parameters during production.

We have used temperature ramps with a pre-heating temperature of 100 °C, and a maximum temperature of 320 °C for 20 seconds, or 340 °C for 10 seconds. The applied pressure during the heat application is also of importance. A rule of thumb is to use 0.2 N per solder bump, which for our ASIC equates to a pressure of 2.8 N.

8.2.2 Bond wire inductance

Wire bonding is perhaps the most widely used technology for packaging of integrated circuits. A gold or aluminium wire is attached to the pad, soldered by high temperature, and the other side of the wire is soldered to the external package.

The inductance of a wire bond can be approximated by the next equation [114]:

$$L = \frac{\mu_0}{2\pi} \cdot l \left[\ln\left(\frac{l}{r} + \sqrt{1 + \frac{l^2}{r^2}} \right) - \sqrt{1 + \frac{r^2}{l^2}} + \frac{r}{l} + \frac{1}{4} \right] \tag{8.1}$$

where μ_0 is the magnetic permeability of free space ($4\pi \times 10^{-7}$ H/m), l is the length of the bond, and r is the diameter of the wire.

The expression can be simplified for $l >> r$ as:

$$L = \frac{\mu_0}{2\pi} \cdot l \left[\ln\left(\frac{2l}{r} \right) - 0.75 \right] \tag{8.2}$$

For example, the inductance of a 1 mm wirebond with diameter of 25 µm is of 0.726 nH, and for a diameter of 40 µm the inductance is of 0.632 nH.

A common rule of thumb for wirebonds of 25 µm diameter is to consider one nanohenry of inductance per millimeter of wirebond, to be on the safe side. Another approximations such as the airbridge inductance are found in [115].

8.2.3 Simulations in HFSS

For flip-chip, the S-parameters of a IHP S13GS pad are extracted, considering a rectangular solder bump made of gold. The pad was drawn on top of a silicon substrate covered with a layer of silicon dioxide. This is shown in Fig. 8.4a. In parallel, the geometry of an ideal wire bond of 25 µm diameter is given in Fig. 8.4b. The wire bond was built using the standard library already present in the software. The specifications are the height, the length and the angles of each segment.

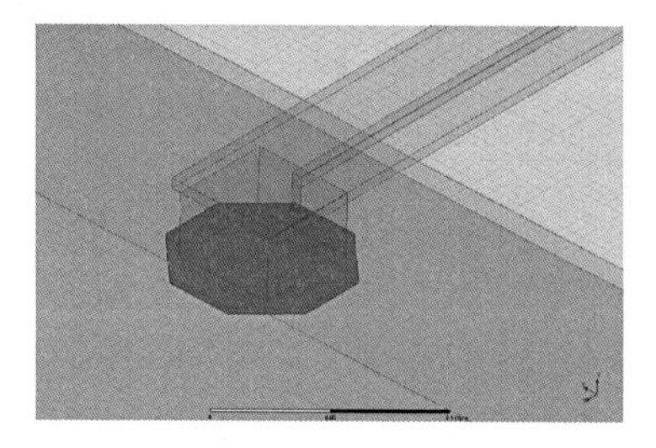
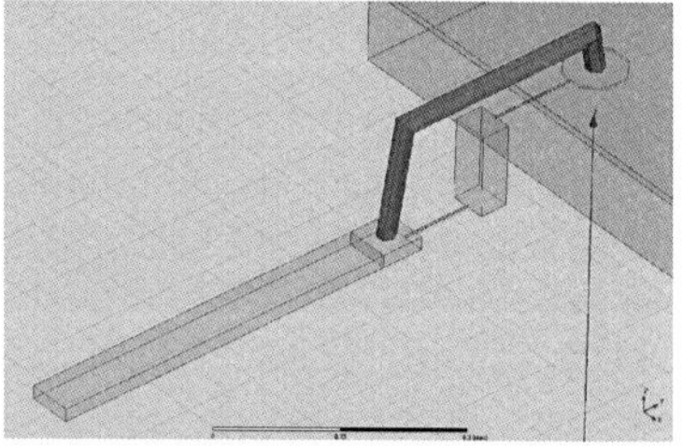

FIGURE 8.4: Model of a flip-chip bump (left) and a wire bond (right) in HFSS.

The S-parameters for both cases were simulated and stored in a s2p file. The curves are shown in Fig. 8.5. Observe that for a wire bond, the reflected power (S11) is higher than the transmitted power (S21) at frequencies above 45 GHz. This does not occur for the flip-chip bump.

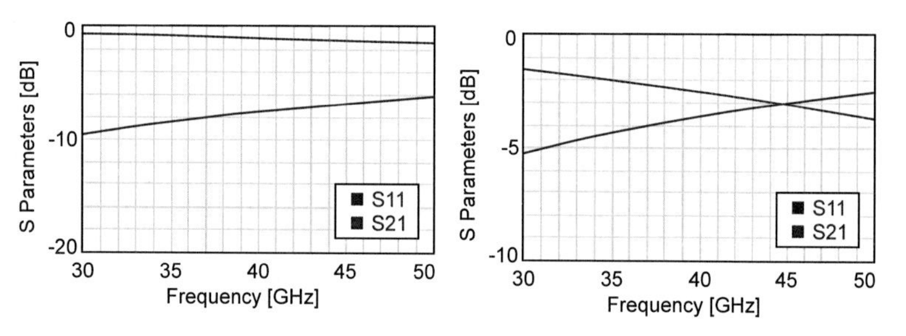

FIGURE 8.5: S-parameters of a flip-chip bump (left) and a wire bond (right).

This is one of the reasons why both technologies have a different limit in the maximum frequency they can support. The assembly of the test board in this chapter is based on wire bonding, and therefore this frequency limitation should be kept in mind.

8.3 Test board assembly

This test board is intended for testing the complete impedance measurement system. The ASIC is glued and wire bonded to a Rogers PCB which includes the connectors for the DC power supplies. On top of this, a second PCB made of alumina is produced with a microstrip line, which is intended to act as the cell sensor. The structure of the test board is shown in Fig. 8.6.

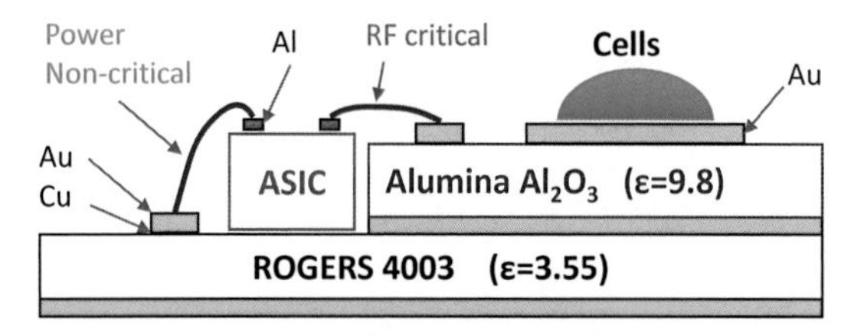

FIGURE 8.6: Stacked PCB for cell experiments up to 50 GHz.

The microstrip line is one type of planar transmission line, the other two being the stripline and the coplanar waveguide. A microstrip line (MLIN) consists of a single strip of metal placed on top a dielectric substrate, with a ground plane attached at the bottom. The placement of two microstrip lines side by side constitutes a coupled microstrip line (MCLIN). The relevant dimensions of a microstrip (MLIN) and a coupled microstrip (MCLIN) are shown in Figure 8.7.

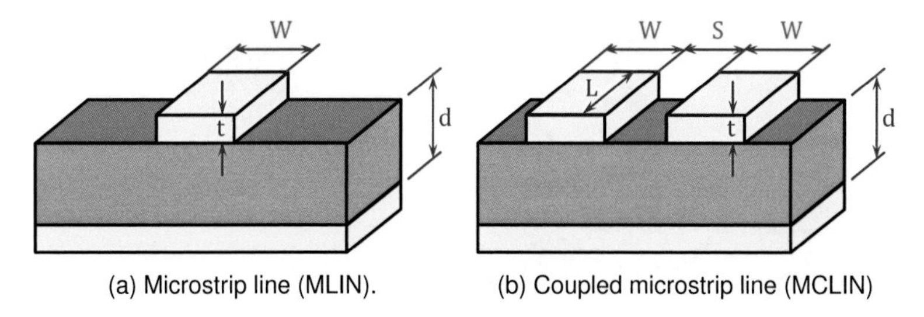

(a) Microstrip line (MLIN).

(b) Coupled microstrip line (MCLIN)

FIGURE 8.7: Dimensions of a microstrip (MLIN) and coupled microstrip line (MCLIN).

The characteristic impedance of the microstrip line (MLIN) can be calculated from empirical relations [116] as follows:

$$Z_0 = \begin{cases} \dfrac{60}{\sqrt{\epsilon_e}} \ln\left(\dfrac{8d}{W} + \dfrac{W}{4d}\right) & (W/d \leq 1) \\ \dfrac{120\pi}{\sqrt{\epsilon_e}[W/d + 1.393 + 0.667\ln(W/d + 1.444)]} & (W/d > 1) \end{cases} \tag{8.3}$$

The effective dielectric constant ϵ_e is:

$$\epsilon_e = \frac{\epsilon_r + 1}{2} + \frac{\epsilon_r - 1}{2} \frac{1}{\sqrt{1 + 12d/W}} \tag{8.4}$$

The characteristic impedance of the coupled microstrip line (MCLIN) may be calculated also from empirical relations, but usually it is more accurate to perform a simulation using specialized design software. We used Linecalc™, which is included in Keysight® ADS™.

The design parameters of the coupled microstrip line are shown in Fig. 8.8. For the alumina PCB we set the permittivity to 9.6, the thickness of the gold layer to 17.5 µm and adjust W and S to be able to obtain a characteristic impedance of 50 Ω.

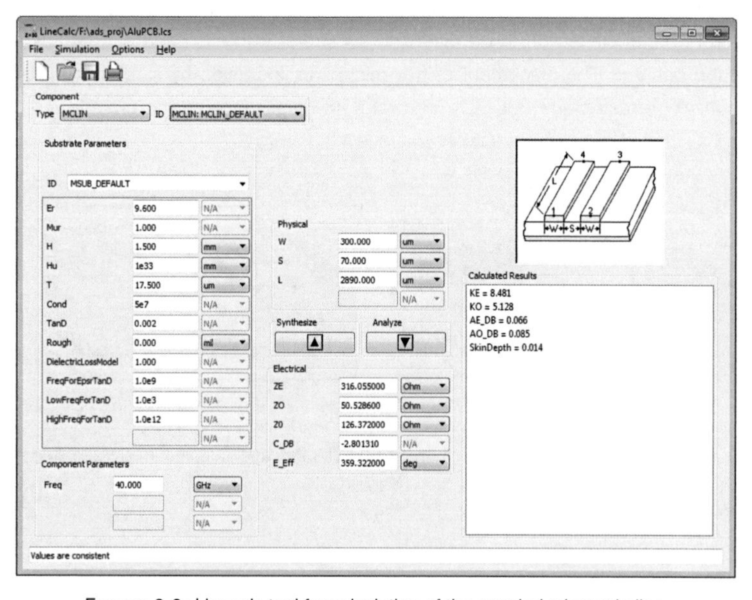

FIGURE 8.8: Linecalc tool for calculation of the coupled microstrip line.

The optimization tool was used to obtain a characteristic impedance of 50 Ω, by adjusting the W and S parameters to 300 μm and 70 μm respectively.

The lenght L of the microstrip line introduces a phase shift, which was designed as a multiple of 180° to minimize the possible reflections from an open termination.

Two finalized PCBs for testing the ASIC 3 are shown in Fig. 8.9.

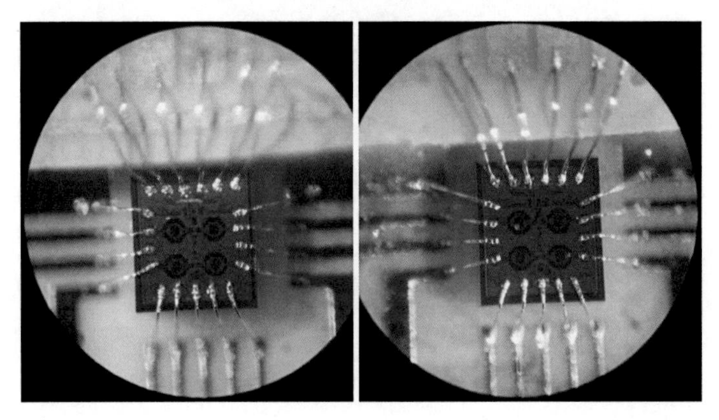

FIGURE 8.9: Two wire bonded dies connected to the stacked PCB structure.

8.4 Oscillator frequency and amplitude

The oscillator frequency and amplitude is determined using a similar setup to the ASIC 2. The following list of equipment was used to perform the measurements:

TABLE 8.2: List of equipment for the oscillator measurements.

Quantity	Description
1	Rhode&Schwarz FSUP50 Signal Source Analyzer, rated 50 GHz.
1	Keithley 4200-SCS Parameter Analyzer for DC power supply.
1	Cascade Microtech Dual InfinityProbe, rated 50 GHz.
5	Karl Suss probe heads with needles for DC power supply.
5	Cascade Microtech DPP105 50 Ω coaxial cable.
1	DC Block INMET 8535K, rated for 40 GHz.
1	50 Ω termination INMET TS400, rated for 40 GHz.
1	Adapter from 2.9 mm to 2.4 mm, rated for 40 GHz.

The measurement of the oscillator with the needles and HF probe is presented in Fig. 8.10. The two needles on the top right seem to be touching each other, but they are not electrically connected. This was confirmed by the electrical measurements. The VCO is oscillating at frequencies between 43.377 GHz, as shown in Fig. 8.11.

FIGURE 8.10: Measurement of the VCO for the ASIC 3.

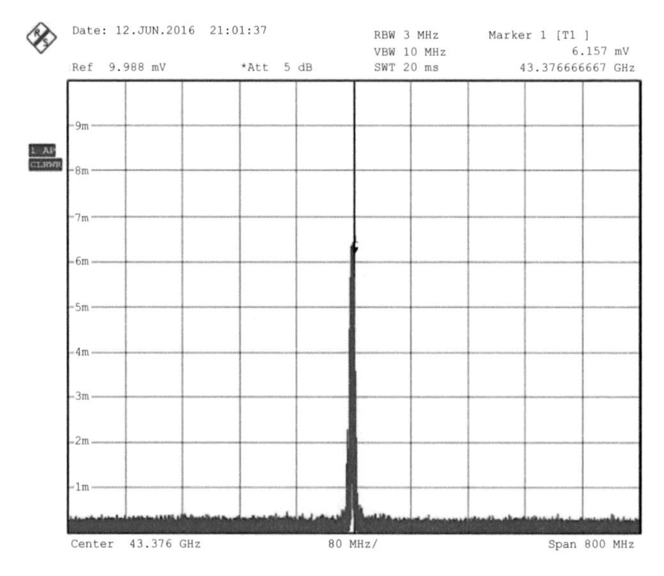

FIGURE 8.11: Output of the votage-controlled oscillator from the ASIC 3.

8.5 DC operating point of the VCO

The measurement results are summarized in Table 8.3. The varactor control voltage V_{ctr} was varied to observe the change in the oscillation frequency from 43.17 GHz to 43.60 GHz. The reference current I_{CM} of the current mirrors was used to vary the output voltage of the oscillator from 6.157 mV up to 8.529 mV.

TABLE 8.3: Results of the oscillator measurement.

Parameter	Simulated	Measured
VCC	3.3 V	3.3 V
GND	0	0
Vctr	0 to 1.5 V	0 to 1.5 V
ICC	25.3 mA	19.92 mA
IC	3mA - 4mA	3mA - 4mA
IGND	25.3 mA	19.92 mA
Freq	38.5 - 46.5 GHz	43.17 – 43.60 GHz
Vout	40 - 60 mV	6.157 mV - 8.529 mV

The power consumption of the ASIC 3 is calculated as the total current multiplied by the supply voltage. This is 3.3 V×19.92 mA=65.736 mW. This is slightly lower than the power consumption of ASIC 2.

8.6 Simulation of an ESD error

In theory, the four outputs of the ASIC 3 should remain at a constant value of approximately 3 V when the load is set to infinite impedance. This was already shown in the simulation from Fig. 8.2, and this condition was measured in earlier experiments with the ASIC.

However, during preliminary measurements of the output voltages, the voltage level for Channel 3 was permanently fixed to 2 V, and the voltage from Channel 4 was tied to a very low level close to 0.2 V. This condition is presented in some of the ASIC samples, and appears after some time of operation, even without external intervention. The circuit may be running and suddenly experience this effect, bringing one channel to ground and the other to 2 V.

The cause of this behavior is the permanent damage of one of the output capacitors. The ASIC is highly sensitive to electrostatic discharges (ESD), since the chip does not include any ESD protection circuitry in the proximity of the pads. The reason for this was not a design mistake, but more of a fundamental problem related to all ASIC designs in the millimeter-wave frequencies.

Typically, ESD protection is implemented by adding diodes and resistors to the pads, in such a way that enable the discharge of these very harmful signals in case of a sudden ESD event. The diodes and resistors add parasitics to the pads, which in most cases are large enough to block the transmission of the high-frequency fields out of the chip. This is one of the reasons why the IHP SG130S design kit does not include any standard cell library for this end, as compared with lower frequency technologies such as the AMS C35B4. The ESD protection pads must be designed carefully for each particular application, and this is a non-trivial task. In addition, the IHP kit does not include ERC checking capabilities.

The damaged capacitor is 37 μm x 71 μm, with a capacitance of 4 pF. The same capacitor is failing systematically in at least two chips. The capacitor is shown in the layout of Fig. 8.12.

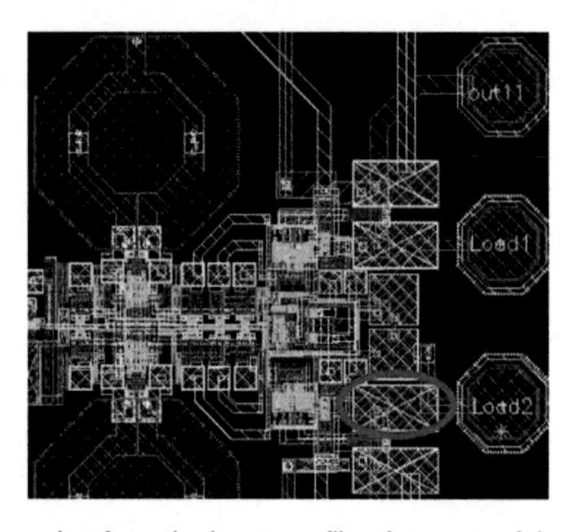

FIGURE 8.12: Capacitor from the low-pass filter from one of the output terminals.

To confirm the observations, the failure of one of the capacitors is simulated by adding a short-circuit across the terminals. The results of this simulation confirm the failure, as it can be seen in Fig. 8.13. This diagram shows that the failure of the capacitor is the cause for the low voltages of two channels.

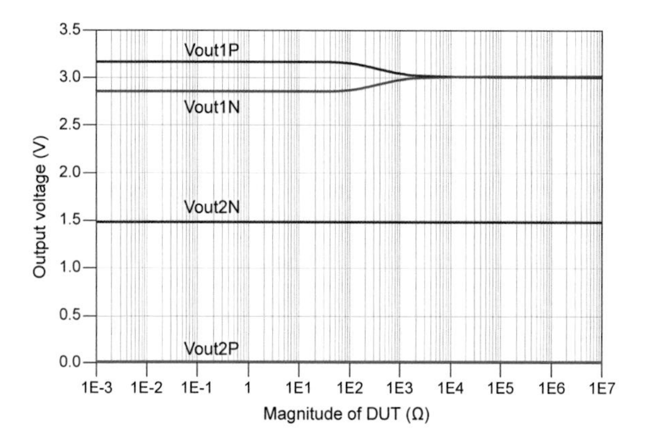

FIGURE 8.13: (a) Simulation of the chip with a short circuit on the LPF capacitor.

8.7 Measurement of cell cultures

In this section we test the measurement of cells by using the cell line CHO-DP12, the acronym expanded as Chinese Hamster Ovary cells. These cells were provided by the cell service laboratory of the Institute of Bioprocess and Biosystems Engineering from the TUHH. For this experiment, a droplet of cells are placed at the end of the microstrip line, enclosed by a rubber ring, as shown in Fig. 8.14.

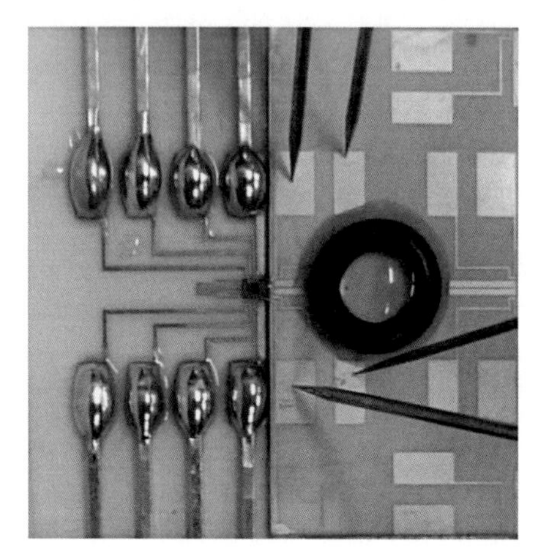

FIGURE 8.14: Cell measurements performed with the ASIC 3.

The results from this measurement are presented in Fig. 8.15.

The figure contents are explained as follows. The lower PCB is connected to the Keithley SCS 4200 which had already been turned on for approximately 15 minutes, but without any power output. At that moment, the DC supply voltages and currents are turned on, which powers on the ASIC. Some minutes later, the cells are deposited at the point marked with a straight vertical line. The ASIC is left running to observe any variation on the output channels.

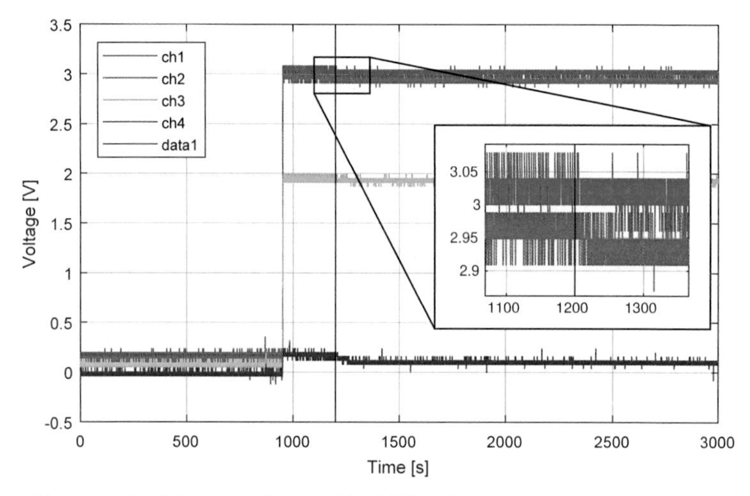

FIGURE 8.15: DC output voltages of the ASIC 3 after depositing a drop of cells, at a time marked with the vertical black line.

The output channels 1 and 2 show a small variation as soon as the cells are deposited, which may be explained by the impedance of the cells. This small variation is not noticeable due to the scale and resolution of the device, but for both channels 1 and 2 it corresponds to at least a voltage difference of 40-50 mV in average, which shows that the cells have a small but noticeable effect in the output channels of the impedance measurement device. This is the proof that the ASIC is working, since it is sensitive to changes in the impedance observed from the load terminals.

The output channels 3 and 4 are tied to values close to 2 V and 0 V, suggesting of permanent ESD damage of one of the output capacitors from the low-pass filters of the ASIC, as it was discussed in the previous section. This was expected and did not prevent observing the small variation of impedance.

8.8 Simulation of the microstrip line

The following simulation models the complete path that the stimulation signal has to travel before being detected by the 16 Ω internal resistor of the ASIC 2 and ASIC 3, since both devices would be connected to a similar circuit for cell experiments.

The oscillator provides the output voltage at the LOAD+ terminal, assuming the chip is properly matched until this point. The signal travels through the first wire bond (Wire1) and then travels to the coupled microstrip line. The cells or the DUT load is placed somewhere in the middle of this configuration, as modeled by the resistance R2. The remaining section of the microstrip line is placed afterwards, terminated by an open circuit. The return path is similar, going back through the microstrip and then entering the ASIC through the second wire bond (Wire2). After this, the internal resistance of 16 Ω is finally reached.

The schematic of the simulation is provided in Fig. 8.16. The oscillator output voltage is assumed to be 1 V for simplification and better interpretation of the results. This voltage is way more than the actual value provided by the oscillator of the ASIC.

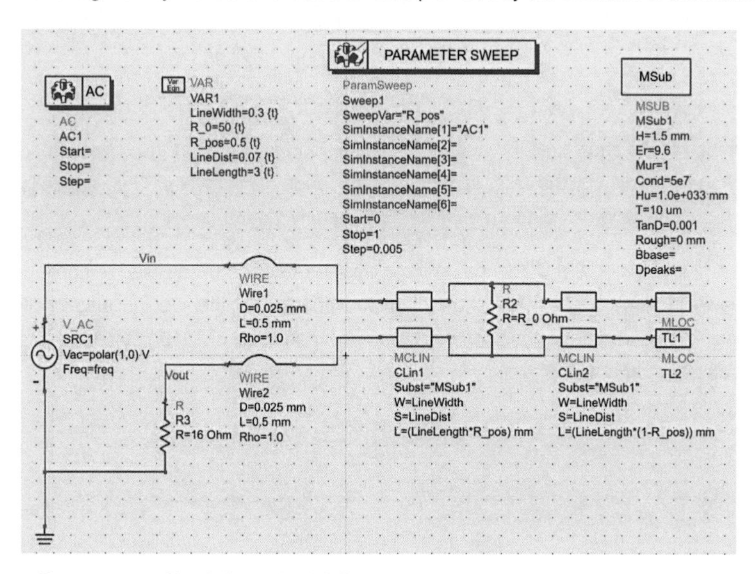

FIGURE 8.16: Simulation of the ASIC, wire bonds, the coupled microstrip line and the internal 16 Ω resistor.

The element R2 is defined in the simulation as a 50 Ω resistor, placed at some position in the middle of the microstrip. In the simulation, the position of this impedance is swept from the beginning (R_pos=0) to the end of the microstrip (R_pos=1), and the results are summarized in Table 8.4. The oscillator voltage is Vin, and the voltage drop across the internal 16 Ω resistor is Vout.

TABLE 8.4: Simulation results obtained by sweeping the position of R2 from the beginning to the end of the microstrip.

Position	Vout [V]	Position	Vout [V]
0.000	8.937 μV	0.500	4.172 μV
0.050	2.098 μV	0.550	12.83 μV
0.100	1.620 μV	0.600	20.56 μV
0.150	7.195 μV	0.650	21.95 μV
0.200	16.31 μV	0.700	16.06 μV
0.250	22.03 μV	0.750	6.959 μV
0.300	20.40 μV	0.800	1.621 μV
0.350	12.56 μV	0.850	2.156 μV
0.400	4.006 μV	0.900	9.194 μV
0.450	1.515 μV	0.950	18.08 μV
		1.000	22.35 μV

These results show that, no matter where the load is placed in the microstrip, the voltage observed by the 16 Ω resistor is in the range of microvolts. This voltage falls in the same range as of thermal noise for the analyzed frequency.

The plot between the measured voltage and the position of the load is nonlinear, shown in Fig. 8.17. This shows that the position of the load is also critical, since there are some positions showing a voltage drop of -116 dBm, equivalent to approximately 1 microvolt.

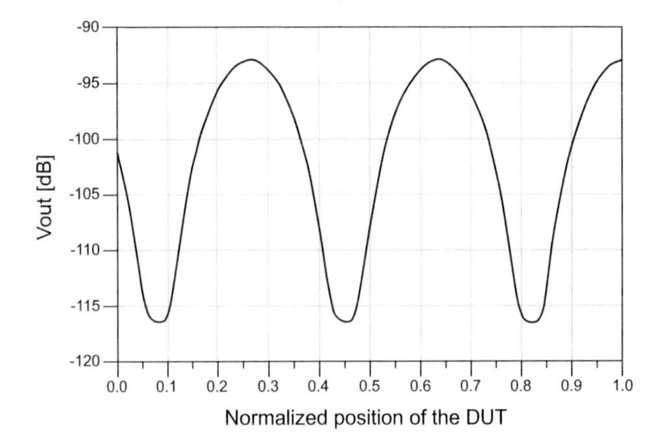

FIGURE 8.17: Voltage observed at the internal 16 Ω resistor as a function of the normalized position of the load (from 0 to 1) on top of the microstrip line.

This simulation highlights the difficulties of connecting external hardware to the ASIC 2 and 3, specifically two wire bonds and a coupled microstrip line which was carefully designed to match the output impedance of the chip. The main problem is the loss of power in the external signal path, which results in a very small voltage drop sensed by the internal 16 Ω resistor. The simulation voltage of 1 V was unrealistic, since the chip is able to provide only 10 mV as measured in the oscillator experiments. The voltage at the internal resistor is then assumed to be two orders of magnitude less, in the 10 nanovolt range, which cannot be detected.

8.9 Chapter conclusion

This chapter describes the test results of the main components of ASIC 3. This includes the oscillator frequency and output voltage measurements, the DC power consumption, and the assembly of a stacked PCB test board for experimentation with cell cultures, where we observed a slight variation in the output signals, of approximately 40-50 mV, after placing a drop of CHO DP-12 cells on top of the microstrip line. This is the proof that the ASIC is sensitive to impedance changes in the surroundings of the microstrip.

The experiments included in this chapter were performed using a custom test fixture with two stacked PCBs. The base circuit board included the connections for the power supplies, and the top board included a couple microstrip line, intended to be used as a cell sensor. The design of this PCB was pursued in collaboration with the Institute of High Frequency. Wire bonding provided solid connectivity for the DC power supply pads, and the circuit operated for long periods without interruption, enabling the measurement of the oscillation frequency. Wire bonding technology required of specialized equipment and we thank Ms. Carmen Hajunga from the Institute of High Frequency for her continuous support with the manufacturing process.

The oscillator of this ASIC is working and producing an oscillation with a frequency of 43.377 GHz, which could be modified in the range from 43.17-43.60 GHz by using the control voltage of the varactors. The amplitude of the oscillation was considerably improved in this redesign, showing an oscillator output voltage which is tunable in the range from 6.157 mV up to 8.529 mV. A detailed comparison of both designs is shown in Table 8.5. Overall, the ASIC 3 is a significant improvement from the previous design.

TABLE 8.5: Comparison of the ASIC 2 and ASIC 3.

Parameter	ASIC 2	ASIC 3
Frequency	38.94-39.65 GHz	43.17-43.60 GHz
Amplitude	0.561-1.780 mV	6.157-8.529 mV
Current consumption	23.37 mA	19.92 mA
Power	77.12 mW	65.74 mW

The test board with the coupled microstrip sensor was used to measure the impedance of a CHO-DP12 cell culture. The output voltages produced a small signal, which was observed as soon as the drop of cells was deposited at the end of the microstrip. This measurement also showed that two output channels were giving a proper response, close to +3 V as simulated, whereas the other two channels showed a DC offset of +2 V and +0.2 V. This was confirmed by the simulation of an ESD error, and it was concluded that one of the output capacitors of the ASIC is getting shorted spontaneously during the measurements. This happens when the chip is powered on, even without any intervention. The same capacitor is failing systematically in more than two ASICs. Although special care was taken during the tests, often the filter capacitors located at the output pads were found to be irreversibly shorted. The design of ESD protection pads for these high-frequency devices is not trivial: it requires detailed planning of specifications and careful design, since at millimiter-wave frequencies, the ESD pads may introduce large parasitics that could not be acceptable for the device operation.

Following this result, a model of the full PCB was developed for simulation. This model considers the wirebonds, the coupled microstrip line, the impedance of the load and the internal 16 Ω resistor included in the ASIC. The oscillator voltage was assumed to be of 1 V, and the detected voltage measured by the internal resistor was only in the range of microvolts. In a more realistic scenario, the oscillator voltage was found to be two orders of magnitude less, which would produce a voltage drop on the sensing resistance in the range of nanovolts, which is already in the thermal voltage level. One possible solution for these problems is to remove the wire bonds and any external circuitry such as the microstrip line from the LOAD terminals of the ASIC. The cells must be placed in direct contact with the chip pads, and this is only possible if the remaining pads are protected by using glob-top.

Overall, this chapter demonstrated that in fact the ASIC 3 is a substantial improvement from the ASIC 2, since the oscillation voltage, frequency and tunable range is higher with the only exception of the frequency control.

A future redesign of the ASIC would need to introduce substantial changes to the topology of the device, removing the 16 Ω internal resistance in favor of a completely symmetrical design, to prevent unbalancing the oscillator. This could be achieved with a transimpedance amplifier, in a similar way as the one used in the ABBM, but designed for the actual frequency range.

The challenges for GHz measurement are still the design of a special chamber for cell placement, a symmetrical impedance measurement system, and a way to connect the ASIC to the external chamber. These three considerations are mandatory for successful experimentation.

Chapter 9

Conclusions and outlook

The microelectronics industry is of the largest growing markets in the world, and the availability of advanced fabrication processes make possible the design and development of custom integrated circuits for biomedical applications. Technologies such as the silicon-germanium process from IHP extend the frequency range of ASICs above 100 GHz. For example, the 130 nm technologies used for the ASIC 2 and ASIC 3 push the f_T to 250 GHz for unity gain, and f_{max} to 340 GHz for the maximum oscillation frequency. The costs of fabrication are high for development, but they are drastically reduced as the medical devices go to mass market.

Miniaturization has given the possibility of integrating advanced medical devices into implants. The present study demonstrated the integration of an impedance measurement device into a surface area of less than one square millimiter (1 mm^2). This is a big advancement considering the size and lack of portability of current commercial devices such as bench LCR meters and potentiostats, which cannot be used easily for a permanent on-body impedance monitoring device such as a smart watch, for example for sports performance measurement during a marathon.

Impedance spectroscopy was found to be an useful and effective technique for a wide range of biological and medical applications studied in Chapter 3. Experiments with fruits, mice tumor differentiation, bacteria identification in infected samples, the preliminary measurement of cell viability, and the experiments for sports intensity assessment showed all positive results. The measurements from this chapter were carried out using two sensing electrodes, either with a potentiostat or with ASICs. The method is generally more accurate when four electrodes are used instead of two. In the four terminal method, the stimulation electrodes (working and counter) are placed apart from the sensing electrodes (working sense and reference). This separate sensing allows better characterization, especially in the low frequencies where the electrochemical double-layer is formed around the stimulation electrodes. A significant advancement would be the integration of four-terminal measurement devices in ASICs for more accurate sensing.

The development of a portable impedance spectroscopy demonstrator in Chapter 5 showed that the auto-balancing bridge method allows recording impedance in both the time and frequency domains. This demonstrator was used to characterize the growth of yeast cells, and also it allows to obtain the frequency response of the samples. The demonstrator gives results with relatively good accuracy, with error percentages below 5% at measurement frequencies of 40 kHz.

In Chapter 6, the assembly of test fixtures for further miniaturization with the ASIC 1 allowed to reduce even more the size, portability and energy consumption of the impedance measurement devices. This integrated circuit measures 16x16 mm and was used to monitor the growth of yeast cells and porcine chondrocytes was observed using the ASIC 1 at frequencies below 40 kHz. For the yeast cells, the presence of sugar accelerated the growth, and it was observed in the impedance response. Porcine chondrocytes on the other hand exhibit a strong decrease after one hour, suggesting the deposition of the cells at the bottom of the measurement chamber, and also giving evidence of cell death. The impedance spectroscopy monitoring is a fast, non-toxic approach which gives instant results without the need of markers or manual counting under the microscope.

The ASIC 2 was used in Chapter 7 for extending the method to the high-frequency range. The circuitry and operation of the device was described, and a simulation example proved that it is possible to calculate the impedance of the DUT from the measurement of the two output voltages of the Gilbert mixers. Later, the DC operating point was measured, as well as the S-parameters of the integrated inductors. The measurements matched the expected results according with simulations in HFSS. Finally, the measurement of the VCO demonstrated that the ASIC is operating in high frequencies. The amplitude of oscillation was found to be small, and possible reasoning for this behavior arises from parasitics and mismatch of the oscillator to the rest of the circuit.

Wire bonding was used in Chapter 8 to assemble a second demonstrator for high-frequency recordings using the ASIC 3. This demonstrator showed considerable improvements when compared to the previous version of the ASIC, in terms of the power consumption, oscillation amplitude and tunable frequency range. The assembly of a stacked PCB allowed measurements of the full impedance measurement system. The device showed some sensitivity to local changes of impedance when a CHO DP-12 cell culture was placed on top of the microstrip line, showing a drop of 40-50 mV in the output channels of the Gilbert mixers. This proves that the ASIC 3 is sensitive to the device-under-test and reacts accordingly.

As a final remark, this work showed that it is possible to integrate complex impedance measurement devices into miniaturized portable systems for biological and medical applications, by integrating ASICs specifically designed for standalone impedance spectroscopy.

Outlook

The present dissertation opens some interesting research fields and design challenges which can be addressed by future projects. A redesign of the ASIC 2 and ASIC 3 would be possible considering a symmetric design where all parts of the circuit are properly matched and balanced, and this would require a full electromagnetic simulation of the chip, interconnection assembly and the experimental chamber with the cells.

In addition, parallelization and simultaneous measurement of multiple cell samples would require special microfluidic devices for contacting the individual compartments, which require careful modeling and simulation to obtain the S-parameters even before the chip design.

The design of custom ASICs could bridge the gap to integration of multiple tools on wearable devices, implants and other critical applications requiring a very optimized device, with low power consumption and standalone operation, featuring portability. The production costs could be reduced when the final products go to mass market commercialization.

The wide range of applications of impedance spectroscopy makes the technique very attractive for further experimentation, since the method could be adapted to virtually any application where the permittivity and conductivity of a material is to be monitored. The portability achieved with this dissertation is a step forward which enables the incursion of this method into new applications.

Appendix A

ASIC bonding plans

A.1 ASIC 1

This ASIC is encapsulated in a PLCC-68 package with the following pin descriptions:

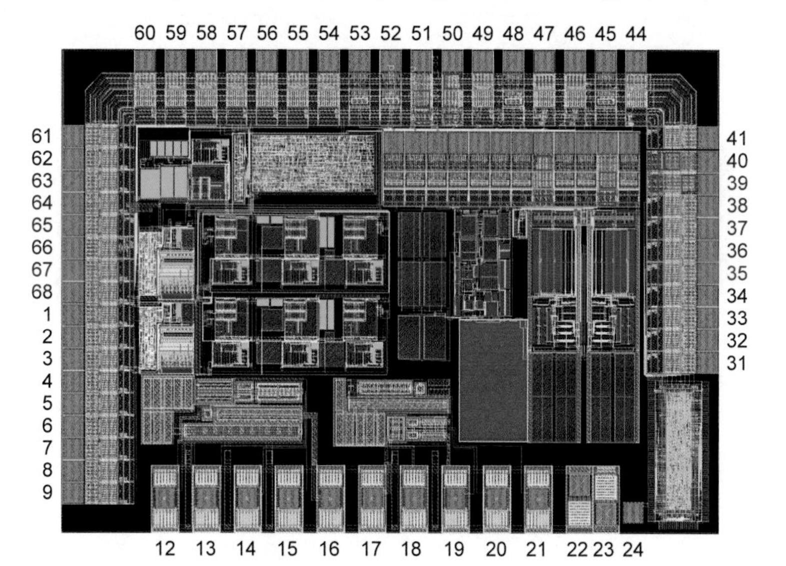

FIGURE A.1: Layout of the ASIC 1.

TABLE A.1: Pin descriptions of the ASIC 1.

Pin number	Pin label	Description
<1:9>	AD1_<8:0>	ADC 1 Digital Output
10	NC	Not connected
11	NC	Not connected
12	I_END	Current mirror for output stage
13	VIN+	kh MillerOTA Input +
14	VIN–	kh MillerOTA Input –
15	I_DIFF	Current mirror for input stage
16	OUT	kh MillerOTA Output
17	OUT1	na MillerOTA Output
18	I1	Current mirror for input stage
19	IOUT	Current mirror for output stage
20	INN	na MillerOTA Input –
21	INP	na MillerOTA Input +
22	VDD+25V	+25V for HV OP-Amps
23	VSS–25	–25V for HV OP-Amps
24	CAP_OUT	Output of capacitor array
<25:30>	NC	Not connected
<31:38>	IW_<0:7>	Address of capacitor array
39	GND_NA	Ground for capacitor array
40	VDD_NA	+3.3 V for capacitor array
41	RSTn	Reset signal
42	NC	Not connected
43	NC	Not connected
44	REGIN	Data pin for 32 bit register
45	EX_OSC	External oscillator pin
46	CLK_REG	Clock pin for 32 bit register
47	CLK_OSC	12.8 MHz clock for internal oscillator
48	SGND	+1.65 V
49	ABBM_OUT	Output of the impedance measurement circuit
50	GND	Ground
51	VDD	+3.3 V
52	OUT2	Device-under-test 2
53	OUT1	Device-under-test 1
54	AD2_EOC	ADC 2 End of Conversion
<55:64>	AD2_<9:0>	ADC 2 Digital Output
65	ADC_START	Enable signal of both ADCs
66	ADC_CLK	Clock signal of both ADCs
67	AD1_EOC	ADC 1 End of Conversion
68	AD1_9	ADC 1 Digital Output

A.2 ASIC 2

The ASIC 2 is provided in bare dies, without encapsulation. The bonding plan is shown in Fig. A.2.

The chip includes flip-chip bumps for the pads of the impedance measurement circuit. Bumping specifications: 80 Au/ 20 Sn, 55 µm bump diameter, 130 µm pitch from PacTech.

The rest of the chip has aluminum pads, and should be measured with needles and high-frequency probes such as the Infinity Probe from Cascade Microtech, part number INF/10017 I50-D-GSGSG-125.

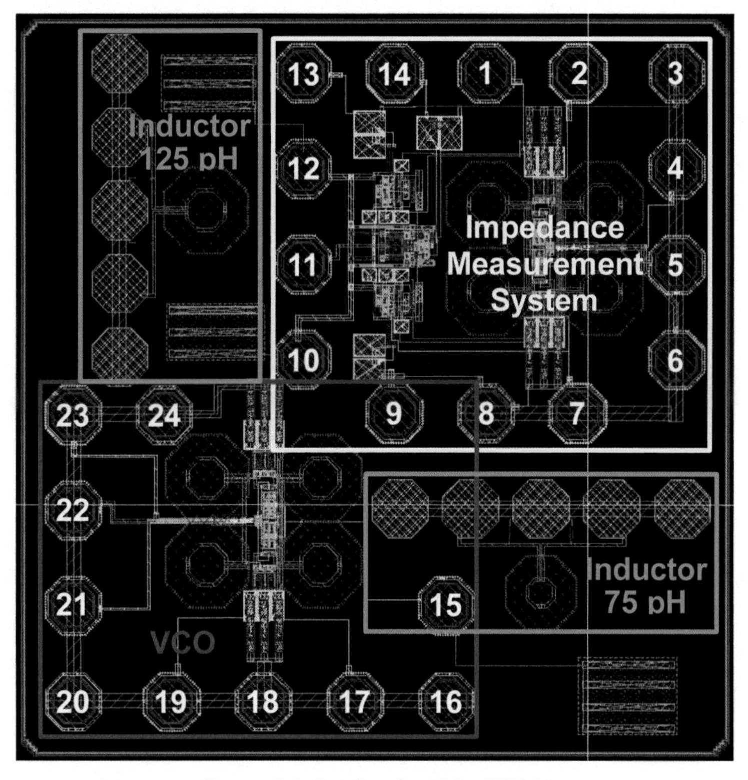

FIGURE A.2: Bonding plan of the ASIC 2.

TABLE A.2: Pin descriptions of the ASIC 2.

Pin number	Pin label	Description
1	LOAD-	Cell container/ device under test
2	LOAD+	Cell container/ device under test
3	VSS	-3.3 V
4	GND	0 V
5	VCONT	Varactor control voltage (-0.45 to -1.65 V)
6	VDD	-1.65V
7	VCM	Reference current of the CM (3-4 mA)
8	VSS	-3.3V
9	GND	0V
10	VSIN	DC output voltage proportional to imag(Z_{DUT})
11	VSS	-3.3V
12	VDD2	0 V
13	VSS	-3.3V
14	VCOS	DC output voltage proportional to real(Z_{DUT})
15	VSS	-3.3 V
16	GND	0 V
17	Port 1	Oscillator output port 1
18	GND	0 V
19	Port 2	Oscillator output port 2
20	GND	0 V
21	VCONT	Varactor control voltage (-0.45 to -1.65 V)
22	VDD	-1.65V
23	VCM	Reference current of the CM (3-4 mA)
24	VSS	-3.3V

Notes:

1. The sample container/device under test must be connected between LOAD- and LOAD+

2. All VSS terminals are connected together and have a voltage of -3.3V. The reason why more than one terminal is used is to distribute the VSS current among them.

3. Both GND terminals are 0V and are connected together inside the integrated circuit. The reason why more than one terminal is used is to distribute the GND current among them.

A.3 ASIC 3

The ASIC 3 is provided in bare dies, without encapsulation or flip-chip bumping. The bonding plan is shown in Fig. A.3.

The left part of the chip is the internal oscillator, and the right part is the complete impedance measurement system including a second oscillator.

The oscillator on the left provides access to the quadrature signals I1 and Q1.

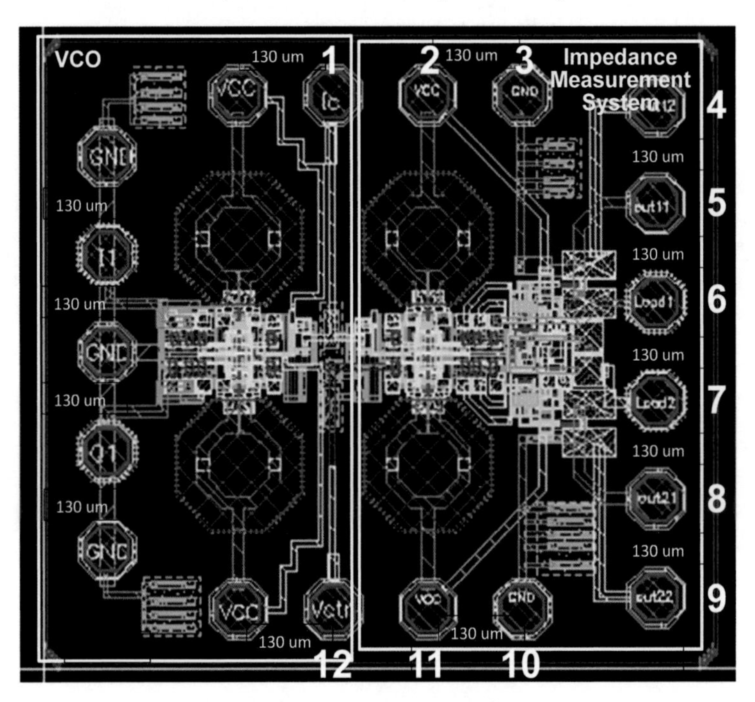

FIGURE A.3: Bonding plan of the ASIC 3.

TABLE A.3: Partial pin descriptions of the ASIC 3.

Pin number	Pin label	Description
1	Ic	Reference current of the CM (3-4 mA)
2	VCC	+3.3 V
3	GND	0 V
4	Out12	DC output voltage proportional to the sine of the angle of the device under test (first output)
5	Out11	DC output voltage proportional to the sine of the angle of the device under test (second output)
6	LOAD-	Cell container/ device under test
7	LOAD+	Cell container/ device under test
8	Out21	DC output voltage proportional to the cosine of the angle of the device under test (first output)
9	Out22	DC output voltage proportional to the cosine of the angle of the device under test (second output)
10	GND	0 V
11	VCC	+3.3 V
12	Vctr	Varactor control voltage (0 to +1.5 V)

Bibliography

[1] Robert Hooke. *Micrographia.* Royal Society, 1665. ISBN 1602066639. doi:10.5962/bhl.title.904.

[2] Eva Bianconi, Allison Piovesan, Federica Facchin, Alina Beraudi, Raffaella Casadei, Flavia Frabetti, Lorenza Vitale, Maria Chiara Pelleri, Simone Tassani, Francesco Piva, Soledad Perez-Amodio, Pierluigi Strippoli, and Silvia Canaider. An estimation of the number of cells in the human body. *Annals of Human Biology*, 40(6):463–471, 2013. ISSN 0301-4460. doi:10.3109/03014460.2013.807878.

[3] Matthew K. Vickaryous and Brian K. Hall. Human cell type diversity, evolution, development, and classification with special reference to cells derived from the neural crest. *Biological Reviews*, 81(03):425, aug 2006. ISSN 1464-7931. doi:10.1017/S1464793106007068.

[4] Rudolf Höber. Eine Methode, die elektrische Leitfähigkeit im Innern von Zellen zu messen. *Pflüger's Archiv für die Gesammte Physiologie des Menschen und der Tiere*, 133(4-6):237–253, 1910. ISSN 00316768. doi:10.1007/BF01680330.

[5] Ursula G. Kyle, Ingvar Bosaeus, Antonio D. De Lorenzo, Paul Deurenberg, Marinos Elia, José Manuel Gómez, Berit Lilienthal Heitmann, Luisa Kent-Smith, Jean-Claude Melchior, Matthias Pirlich, Hermann Scharfetter, Annemie M.W.J. Schols, and Claude Pichard. Bioelectrical impedance analysis—part I: review of principles and methods. *Clinical Nutrition*, 23(5):1226 – 1243, 2004. ISSN 0261-5614. doi:10.1016/j.clnu.2004.06.004.

[6] A. Hickling. Studies in electrode polarisation. Part I. The accurate measurement of the potential of a polarised electrode. *Transactions of the Faraday Society*, 33: 1540–1546, 1937. ISSN 00147672. doi:10.1039/TF9373301540.

[7] A. Hickling. Studies in electrode polarisation. Part II. The investigation of the rate of growth of polarisation potentials. *Transactions of the Faraday Society*, 35:364, 1940. ISSN 0014-7672. doi:10.1039/tf9403500364.

[8] A. Hickling. Studies in electrode polarisation. Part III. A note on the theoretical implications of the rate of growth of polarisation potentials. *Transactions of the Faraday Society*, 37(532):532, 1941. ISSN 0014-7672. doi:10.1039/tf9413700532.

[9] A. Hickling. Studies in electrode polarisation. Part IV.—The automatic control of the potential of a working electrode. *Trans. Faraday Soc.*, 38:27–33, 1942. ISSN 0014-7672. doi:10.1039/TF9423800027.

[10] K. Jüttner. Electrochemical impedance spectroscopy (EIS) of corrosion processes on inhomogeneous surfaces. *Electrochimica Acta*, 35(10):1501–1508, 1990. ISSN 00134686. doi:10.1016/0013-4686(90)80004-8.

[11] Xiao-Zi Yuan, Chaojie Song, Haijiang Wang, and Jiujun Zhang. *Electrochemical Impedance Spectroscopy in PEM Fuel Cells*. Springer London, London, 2010. ISBN 978-1-84882-845-2. doi:10.1007/978-1-84882-846-9.

[12] A. Soley, M. Lecina, X. Gámez, J.J. Cairó, P. Riu, X. Rosell, R. Bragós, and F. Gòdia. On-line monitoring of yeast cell growth by impedance spectroscopy. *Journal of Biotechnology*, 118(4):398–405, sep 2005. ISSN 01681656. doi:10.1016/j.jbiotec.2005.05.022.

[13] A. Ur and D. F. J. Brown. Impedance Monitoring of Bacterial Activity. *Journal of Medical Microbiology*, 8(1):19–28, feb 1975. ISSN 0022-2615. doi:10.1099/00222615-8-1-19.

[14] J C S Richards, A. C. Jason, G. Hobbs, D. M. Gibson, and R. H. Christie. Electronic measurement of bacterial growth. *Journal of Physics E: Scientific Instruments*, 11(6): 560–568, jun 1978. ISSN 0022-3735. doi:10.1088/0022-3735/11/6/017.

[15] A. Chowdhury, T. K. Bera, D. Ghoshal, and B. Chakraborty. Studying the electrical impedance variations in banana ripening using electrical impedance spectroscopy (EIS). *Proceedings of the 2015 3rd International Conference on Computer, Communication, Control and Information Technology, C3IT 2015*, pages 1–4, 2015. doi:10.1109/C3IT.2015.7060196.

[16] Mahfoozur Rehman, Basem A.J.A. Abu Izneid, Mohd Zaid Abdullah, and Mohd Rizal Arshad. Assessment of quality of fruits using impedance spectroscopy. *International Journal of Food Science & Technology*, 46(6):1303–1309, 2011. ISSN 09505423. doi:10.1111/j.1365-2621.2011.02636.x.

[17] Acácio Figueiredo Neto, Nelson Cárdenas Olivier, Erlon Rabelo Cordeiro, and Helinando Pequeno de Oliveira. Determination of mango ripening degree by electrical impedance spectroscopy. *Computers and Electronics in Agriculture*, 143(September): 222–226, 2017. ISSN 01681699. doi:10.1016/j.compag.2017.10.018.

[18] Anne D Bauchot, F.Roger Harker, and W.Michael Arnold. The use of electrical impedance spectroscopy to assess the physiological condition of kiwifruit. *Postharvest Biology and Technology*, 18(1):9–18, 2000. ISSN 09255214. doi:10.1016/S0925-5214(99)00056-3.

[19] Laura Ceriotti, Jessica Ponti, Pascal Colpo, Enrico Sabbioni, and Francois Rossi. Assessment of cytotoxicity by impedance spectroscopy. *Biosensors and Bioelectronics*, 22(12):3057–3063, jun 2007. ISSN 09565663. doi:10.1016/j.bios.2007.01.004.

[20] A. Caduff, F. Dewarrat, M. Talary, G. Stalder, L. Heinemann, and Yu. Feldman. Non-invasive glucose monitoring in patients with diabetes: A novel system based on impedance spectroscopy. *Biosensors and Bioelectronics*, 22(5):598–604, dec 2006. ISSN 09565663. doi:10.1016/j.bios.2006.01.031.

[21] R L Alvarenga and M N Souza. Estimation of the Lactate Threshold Using Bioelectrical Impedance Spectroscopy: A New Noninvasive Method. In *2007 29th Annual International Conference of the IEEE Engineering in Medicine and Biology Society*, volume 2007, pages 3052–3055. IEEE, aug 2007. ISBN 978-1-4244-0787-3. doi:10.1109/IEMBS.2007.4352972.

[22] R. Anderson and O. Dennison. An advanced new network analyzer for sweep-measuring amplitude and phase from 0.1 to 12.4 GHz. *Hewlett Packard J.*, 18(1):2–9, 1967.

[23] Margaret Cheney, David Isaacson, and Jonathan C. Newell. Electrical Impedance Tomography. *SIAM Review*, 41(1):85–101, 1999. ISSN 0036-1445. doi:10.1137/S0036144598333613.

[24] Ann P O'Rourke, Mariya Lazebnik, John M Bertram, Mark C Converse, Susan C Hagness, John G Webster, and David M Mahvi. Dielectric properties of human normal, malignant and cirrhotic liver tissue: in vivo and ex vivo measurements from 0.5 to 20 GHz using a precision open-ended coaxial probe. *Physics in Medicine and Biology*, 52(15):4707–4719, aug 2007. ISSN 0031-9155. doi:10.1088/0031-9155/52/15/022.

[25] L. A. Geddes and H. E. Hoff. The discovery of bioelectricity and current electricity the Galvani-Volta controversy. *IEEE Spectrum*, 8(12):38–46, 1971. ISSN 00189235. doi:10.1109/MSPEC.1971.5217888.

[26] Emil Du Bois-Reymond. *Untersuchungen über thierische elektricität.* G. Reimer, 1848.

[27] Nicola A. Maffiuletti, Marco A. Minetto, Dario Farina, and Roberto Bottinelli. Electrical stimulation for neuromuscular testing and training: State-of-the art and unresolved issues. *European Journal of Applied Physiology*, 111(10):2391–2397, 2011. ISSN 14396319. doi:10.1007/s00421-011-2133-7.

[28] A. L. Hodgkin. The Ionic Basis of Nervous Conduction. *Science*, 145(3638): 1287–1287, sep 1964. ISSN 0036-8075. doi:10.1126/science.145.3637.1148.

[29] A. F. Huxley. Excitation and Conduction in Nerve: Quantitative Analysis. *Science*, 145 (3637):1154–1159, sep 1964. ISSN 0036-8075. doi:10.1126/science.145.3637.1154.

[30] Erwin Neher and Bert Sakmann. Single-channel currents recorded from membrane of denervated frog muscle fibres. *Nature*, 260(5554):799–802, apr 1976. ISSN 0028-0836. doi:10.1038/260799a0.

[31] Erwin Neher. Ion channels for communication between and within cells. *Bioscience Reports*, 12(1):1–14, feb 1992. ISSN 0144-8463. doi:10.1007/BF01125822.

[32] Bert Sakmann. Elementary steps in synaptic transmission revealed by currents through single ion channels. *Science*, 256(5056):503–512, apr 1992. ISSN 0036-8075. doi:10.1126/science.1373907.

[33] Jan Gimsa. A comprehensive approach to electro-orientation, electrodeformation, dielectrophoresis, and electrorotation of ellipsoidal particles and biological cells. *Bioelectrochemistry*, 54:23–31, 2001.

[34] Joseph Mangano and Henry Eppich. Cell separation using electric fields, 2003.

[35] Bing Song, Yu Gu, Jin Pu, Brian Reid, Zhiqiang Zhao, and Min Zhao. Application of direct current electric fields to cells and tissues in vitro and modulation of wound electric field in vivo. *Nature Protocols*, 2(6):1479–1489, jun 2007. ISSN 1754-2189. doi:10.1038/nprot.2007.205.

[36] GregoryJ Della Rocca. The science of ultrasound therapy for fracture healing. *Indian Journal of Orthopaedics*, 43(2):121, 2009. ISSN 0019-5413. doi:10.4103/0019-5413.50845.

[37] J A Spadaro. Electrically stimulated bone growth in animals and man. Review of the literature. *Clin Orthop Relat Res*, pages 325–332, 1977. ISSN 0009-921X (Print) 0009-921x.

[38] Luís Ni–o De Rivera, Ernesto Paredes Martínez, Daniel Robles Camarillo, and Wilfrido Calleja Arriaga. Adaptive Electrical Stimulation to Improve In-Vitro Cell Growth. *Procedia Technology*, 3:316–323, 2012. ISSN 22120173. doi:10.1016/j.protcy.2012.03.034.

[39] Keun A. Chang, Jin Won Kim, Jeong a. Kim, Sungeun Lee, Saeromi Kim, Won Hyuk Suh, Hye Sun Kim, Sunghoon Kwon, Sung June Kim, and Yoo Hun Suh. Biphasic electrical currents stimulation promotes both proliferation and differentiation of fetal neural stem cells. *PLoS ONE*, 6(4), 2011. ISSN 19326203. doi:10.1371/journal.pone.0018738.

[40] Giuseppina Covello, Kavitha Siva, Valentina Adami, and Michela A. Denti. An electroporation protocol for efficient DNA transfection in PC12 cells. *Cytotechnology*, 66(4):543–553, aug 2014. ISSN 0920-9069. doi:10.1007/s10616-013-9608-9.

[41] Invitrogen. Neon ™ Transfection System. User Guide. Technical report, Invitrogen, 2010.

[42] Bio-Rad. Gene Pulser Xcell Electroporation System. Instruction Manual. Technical report, Bio-Rad, 2006.

[43] Eilon D Kirson, Vladimír Dbalý, Frantisek Tovarys, Josef Vymazal, Jean F Soustiel, Aviran Itzhaki, Daniel Mordechovich, Shirley Steinberg-Shapira, Zoya Gurvich, Rosa Schneiderman, Yoram Wasserman, Marc Salzberg, Bernhard Ryffel, Dorit Goldsher, Erez Dekel, and Yoram Palti. Alternating electric fields arrest cell proliferation in animal tumor models and human brain tumors. *Proceedings of the National Academy of Sciences of the United States of America*, 104(24):10152–10157, 2007. ISSN 0027-8424. doi:10.1073/pnas.0702916104.

[44] E. D. Kirson. Disruption of Cancer Cell Replication by Alternating Electric Fields. *Cancer Research*, 64(9):3288–3295, may 2004. ISSN 0008-5472. doi:10.1158/0008-5472.CAN-04-0083.

[45] G.E. Moore. Cramming More Components Onto Integrated Circuits. *Proceedings of the IEEE*, 86(1):82–85, jan 1998. ISSN 0018-9219. doi:10.1109/JPROC.1998.658762.

[46] Juan J. Montero-Rodríguez, Edgar Eduardo Salazar-Flórez, Paola Vega-Castillo, Jakob M. Tomasik, Wjatscheslaw Galjan, Kristian M. Hafkemeyer, and Wolfgang Krautschneider. An Impedance Spectroscopy ASIC for Low-Frequency Characterization of Biological Samples. In *Proceedings of the 9th International Joint Conference on Biomedical Engineering Systems and Technologies (BIOSTEC 2016) - Volume 1: BIODEVICES*, pages 222–228, 2016.

[47] Juan J. Montero-Rodríguez, A.J. Fernández-Castro, D. Schroeder, and W. Krautschneider. Development of an impedance spectroscopy device for on-line cell growth monitoring. *Electronics Letters*, 53(15):1025–1027, July 2017. ISSN 0013-5194. doi:10.1049/el.2017.0390.

[48] H.P. Schwan. Electrical properties of tissues and cell suspensions: mechanisms and models. In *Proceedings of 16th Annual International Conference of the IEEE Engineering in Medicine and Biology Society*, pages A70–A71. IEEE, 1994. ISBN 0-7803-2050-6. doi:10.1109/IEMBS.1994.412155.

[49] Herman P. Schwan. Electrical properties of tissue and cell suspensions. *Advances in Biological and Medical Physics*, 5:147–209, 1957. doi:10.1016/B978-1-4832-3111-2.50008-0.

[50] Kenneth S. Cole. *Membranes, Ions and Impulses.* University of California Press, 1972. ISBN 9780520002517.

[51] Allen J Bard and Larry R Faulkner. *Electrochemical Methods - Fundamentals and Applications.* John Wiley & Sons, Inc., 1944. ISBN 9780123813749. doi:10.1016/B978-0-12-381373-2.00056-9.

[52] Hainan Wang and Laurent Pilon. Accurate simulations of electric double layer capacitance of ultramicroelectrodes. *Journal of Physical Chemistry C*, 115(33): 16711–16719, 2011. ISSN 19327447. doi:10.1021/jp204498e.

[53] Reto B. Schoch, Harald van Lintel, and Philippe Renaud. Effect of the surface charge on ion transport through nanoslits. *Physics of Fluids*, 17(10):100604, 2005. ISSN 10706631. doi:10.1063/1.1896936.

[54] Mohammad A. Alim. *Immittance Spectroscopy: Applications to Material Systems.* John Wiley & Sons, Inc., Hoboken, NJ, USA, dec 2017. ISBN 9781119185413. doi:10.1002/9781119185413.

[55] J. E. B. Randles. Kinetics of rapid electrode reactions. *Discussions of the Faraday Society*, 1(4):11, apr 1947. ISSN 0366-9033. doi:10.1039/df9470100011.

[56] a H Kyle, C T Chan, and a I Minchinton. Characterization of three-dimensional tissue cultures using electrical impedance spectroscopy. *Biophysical journal*, 76(5): 2640–2648, 1999. ISSN 0006-3495. doi:10.1016/S0006-3495(99)77416-3.

[57] Min-Haw Wang, Min-Feng Kao, Haw-Juin Liu, Wai-Hong Kan, Yi-Chu Hsu, and Ling-Sheng Jang. A Microfluidic Device for Capture of Single Cells and Impedance Measurement. In *2007 2nd IEEE International Conference on Nano/Micro Engineered and Molecular Systems*, volume 9, pages 711–714. IEEE, jan 2007. ISBN 1-4244-0609-9. doi:10.1109/NEMS.2007.352117.

[58] Min-haw Wang and Wen-hao Chang. Effect of Electrode Shape on Impedance of Single HeLa Cell: A COMSOL Simulation. *BioMed Research International*, 2015: 1–9, 2015. ISSN 2314-6133. doi:10.1155/2015/871603.

[59] Hywel Morgan, Tao Sun, David Holmes, Shady Gawad, and Nicolas G Green. Single cell dielectric spectroscopy. *Journal of Physics D: Applied Physics*, 40(1):61–70, jan 2007. ISSN 0022-3727. doi:10.1088/0022-3727/40/1/S10.

[60] David Holmes, Tao Sun, Hywel Morgan, Judith Holloway, Julie Cakebread, and Donna Davies. Label-free differential leukocyte counts using a microfabricated, single-cell impedance spectrometer. *Proceedings of IEEE Sensors*, pages 1452–1455, 2007. ISSN 1930-0395. doi:10.1109/ICSENS.2007.4388687.

[61] Muhammad Mansor, Masaru Takeuchi, Masahiro Nakajima, Yasuhisa Hasegawa, and Mohd Ahmad. Electrical Impedance Spectroscopy for Detection of Cells in Suspensions Using Microfluidic Device with Integrated Microneedles. *Applied Sciences*, 7(2):170, 2017. ISSN 2076-3417. doi:10.3390/app7020170.

[62] K.K. Karkkainen, A.H. Sihvola, and K.I. Nikoskinen. Effective permittivity of mixtures: numerical validation by the FDTD method. *IEEE Transactions on Geoscience and Remote Sensing*, 38(3):1303–1308, may 2000. ISSN 01962892. doi:10.1109/36.843023.

[63] Yulia Polevaya, Irina Ermolina, Michael Schlesinger, Ben-Zion Ginzburg, and Yuri Feldman. Time domain dielectric spectroscopy study of human cells. *Biochimica et Biophysica Acta (BBA) - Biomembranes*, 1419(2):257–271, jul 1999. ISSN 00052736. doi:10.1016/S0005-2736(99)00072-3.

[64] Tetsuya Hanai. Theory of the dielectric dispersion due to the interfacial polarization and its application to emulsions. *Kolloid-Zeitschrift*, 171(1):23–31, jul 1960. ISSN 0303-402X. doi:10.1007/BF01520320.

[65] H Kaneko, K Asami, and T Hanai. Dielectric analysis of sheep erythrocyte ghost. Examination of applicability of dielectric mixture equations. *Colloid & Polymer Science*, 269(10):1039–1044, oct 1991. ISSN 0303-402X. doi:10.1007/BF00657434.

[66] Edgar Eduardo Salazar Flórez. Simulation and Characterization of Biological Samples Using Impedance Spectroscopy. Master's thesis, Hamburg University of Technology, 2015.

[67] I Ermolina, Yu. Polevaya, and Yu. Feldman. Analysis of dielectric spectra of eukaryotic cells by computer modeling. *European Biophysics Journal*, 29(2):141–145, may 2000. ISSN 0175-7571. doi:10.1007/s002490050259.

[68] Jean-Luc Dellis. Zfit: function which can plot, simulate and fit impedance data, 2010. URL http://www.mathworks.com/matlabcentral/fileexchange/19460-zfit.

[69] J. Ross Macdonald and Larry D. Potter. A flexible procedure for analyzing impedance spectroscopy results: Description and illustrations. *Solid State Ionics*, 24(1):61–79, 1987. ISSN 01672738. doi:10.1016/0167-2738(87)90068-3.

[70] J. Ross Macdonald. LEVM/LEVMW Manual: Immitance, Inversion, and Simulation Fitting Programs for WINDOWS and MS-DOS. Technical report, University of North Carolina, 2015. URL http://jrossmacdonald.com/levmlevmw/.

[71] Soren Koch. Elchemea Analytical. Technical report, Danmarks Tekniske Universitet. Institut for Energikonvertering og -lagring, 2017. URL http://www.elchemea.com/.

[72] G.A. Ragoisha and A.S. Bondarenko. Potentiodynamic electrochemical impedance spectroscopy. *Electrochimica Acta*, 50(7-8):1553–1563, feb 2005. ISSN 00134686. doi:10.1016/j.electacta.2004.10.055.

[73] Alexander S. Bondarenko and Genady A. Ragoisha. EIS Spectrum Analyser: analysis and simulation of impedance spectra, 2013. URL http://www.abc.chemistry.bsu.by/vi/analyser/.

[74] Gamry Inc. Echem Analyst Software Manual. Technical report, Gamry Inc., 2011. URL https://www.gamry.com/support/software/.

[75] Metrohm Autolab. NOVA User manual. Technical report, Metrohm Autolab, 2016. URL http://www.metrohm-autolab.com/.

[76] BioLogic. EC-Lab Software: Techniques and Applications. Technical Report August, Bio-Logic Science Instruments, 2014. URL http://www.bio-logic.net/en/divisions/ec-lab/.

[77] RHD Instruments. RelaxIS Impedance Spectrum Analysis - User's manual. Technical report, RHD Instruments, 2017. URL https://www.rhd-instruments.de.

[78] Kumho Petrochemical Co. Multiple Electrochemical Impedance Spectra Parameterization - User Manual. Technical report, Powergraphy, 2002.

[79] AMETEK Inc. ZSimpWin, 2015. URL http://www.ameteksi.com/products/software/zsimpwin.

[80] Bernard A. Boukamp. A nonlinear least squares fit procedure for analysis of immittancs data of electrochemical systems. *Solid State Ionics*, 20:31–44, 1986. ISSN 01672738. doi:10.1016/0167-2738(86)90031-7.

[81] Bernard A. Boukamp. A package for impedance/admittance data analysis. *Solid State Ionics*, 18-19:136–140, jan 1986. ISSN 01672738. doi:10.1016/0167-2738(86)90100-1.

[82] Novocontrol Technologies. WinFIT Curve Fitting Software For Non-linear Models, 2017. URL http://www.novocontrol.de/php/winfit.php.

[83] Zivelab. Zman user's manual. Technical report, Zivelab, 2014. URL http://zivelab.com/.

[84] Ivium Technologies. IviumSoft Manual. Technical Report June, Ivium Technologies, 2016. URL http://www.ivium.nl.

[85] Scribner Associates Inc. ZView: Impedance / gain phase graphing and analysis software. Operation manual. Technical report, Scribner Associates Inc., 2016. URL http://www.scribner.com/.

[86] Donald W. Marquardt. An Algorithm for Least-Squares Estimation of Nonlinear Parameters. *Journal of the Society for Industrial and Applied Mathematics*, 11(2): 431–441, jun 1963. ISSN 0368-4245. doi:10.1137/0111030.

[87] J. A. Nelder and R. Mead. A Simplex Method for Function Minimization. *The Computer Journal*, 7(4):308–313, 1965. ISSN 0010-4620. doi:10.1093/comjnl/7.4.308.

[88] E. Farhi, Y. Debab, and P. Willendrup. IFit: A new data analysis framework. Applications for data reduction and optimization of neutron scattering instrument simulations with McStas. In *Journal of Neutron Research*, volume 17, pages 5–18, 2014. doi:10.3233/JNR-130001.

[89] John D'Errico. Bound constrained optimization using fminsearch, 2012. URL https://www.mathworks.com/matlabcentral/fileexchange/8277-fminsearchbnd--fminsearchcon.

[90] Andrzej Lasia. *Electrochemical Impedance Spectroscopy and its Applications*. Springer, 1st edition, 2014.

[91] MathWorks. Linear regression, 2016. URL https://de.mathworks.com/help/matlab/data_analysis/linear-regression.html.

[92] Michael R. Stratton, Peter J. Campbell, and P. Andrew Futreal. The cancer genome. *Nature*, 458(7239):719–724, 2009. ISSN 00280836. doi:10.1038/nature07943.

[93] World Health Organization. *International Classification of Diseases for Oncology (ICD-O-3)*. World Health Organization, Geneva, 3rd edition, 2013. URL http://codes.iarc.fr/home.

[94] Ansgar Malich, Thomas Böhm, Tobias Fritsch, Miriam Facius, Martin G. Freesmeyer, Roselle Anderson, Marlies Fleck, and Werner A. Kaiser. Animal-based model to investigate the minimum tumor size detectable with an electrical impedance

scanning technique. *Academic Radiology*, 10(1):37–44, 2003. ISSN 10766332. doi:10.1016/S1076-6332(03)80786-9.

[95] J H Seppenwoolde, P R Seevinck, J F W Nijsen, T C De Wit, F J Beekman, a D Van Het Schip, and C J G Bakker. Sensitivity and detection limits of MRI, CT and SPECT for Holmium-loaded microspheres. *International Society Magnetic Resonance in Medicine*, 313(January 2016):3511, 2006.

[96] José E Belizário. Immunodeficient Mouse Models: An Overview. *The Open Immunology Journal*, 2:79–85, 2009. ISSN 18742262. doi:10.2174/1874226200902010079.

[97] Imke Müller and Sebastian Ullrich. The PFP/RAG2 Double-Knockout Mouse in Metastasis Research: Small-Cell Lung Cancer and Prostate Cancer. In *Methods in Molecular Biology*, volume 1070, pages 191–201. Springer, 2014. ISBN 9781461482437. doi:10.1007/978-1-4614-8244-4_14.

[98] Christopher L Morton and Peter J Houghton. Establishment of human tumor xenografts in immunodeficient mice. *Nature Protocols*, 2(2):247–250, 2007. ISSN 1750-2799. doi:10.1038/nprot.2007.25.

[99] V. Mishra, H. Bouayad, A. Schned, J. Heaney, and R. J. Halter. Electrical impedance spectroscopy for prostate cancer diagnosis. *2012 Annual International Conference of the IEEE Engineering in Medicine and Biology Society*, 1(111):3258–3261, 2012. doi:10.1109/EMBC.2012.6346660.

[100] R. J. Halter, A. Hartov, J. A. Heaney, K. D. Paulsen, and A. R. Schned. Electrical Impedance Spectroscopy of the Human Prostate. *IEEE Transactions on Biomedical Engineering*, 54(7):1321–1327, 2007. ISSN 0018-9294. doi:10.1109/TBME.2007.897331.

[101] Gary J Saulnier, Ning Liu, Chandana Tamma, Hongjun Xia, Tzu-jen Kao, J C Newell, and David Isaacson. An Electrical Impedance Spectroscopy System for Breast Cancer Detection. *IEEE Conf.*, 4:4154–4157, 2007. ISSN 1557-170X. doi:10.1109/IEMBS.2007.4353251.

[102] P. Aberg, I. Nicancer, and S. Ollmar. Minimally invasive electrical impedance spectroscopy of skin exemplified by skin cancer assessments. *Proceedings of the 25th Annual International Conference of the IEEE Engineering in Medicine and Biology Society (IEEE Cat. No.03CH37439)*, pages 3211–3214, 2003. ISSN 1094687X. doi:10.1109/IEMBS.2003.1280826.

[103] Jan-Patrick Kalckhoff. Ionic Experimentation, Modelling and Simulation for Cell Solution Characterization. Master's thesis, Hamburg University of Technology, 2018.

[104] Wei Gao, Sam Emaminejad, Hnin Yin Yin Nyein, Samyuktha Challa, Kevin Chen, Austin Peck, Hossain M. Fahad, Hiroki Ota, Hiroshi Shiraki, Daisuke Kiriya, Der-Hsien Lien, George A. Brooks, Ronald W. Davis, and Ali Javey. Fully integrated wearable sensor arrays for multiplexed in situ perspiration analysis. *Nature*, 529(7587): 509–514, jan 2016. ISSN 0028-0836. doi:10.1038/nature16521.

[105] S. Tornroth-Horsefield and R. Neutze. Opening and closing the metabolite gate. *Proceedings of the National Academy of Sciences*, 105(50):19565–19566, 2008. ISSN 0027-8424. doi:10.1073/pnas.0810654106.

[106] Stryer L Berg JM, Tymoczko JL. *Biochemistry*, chapter Section 18.6: The Regulation of Cellular Respiration Is Governed Primarily by the Need for ATP. Palgrave Macmillan, New York, 5th edition edition, 2002. Available from: https://www.ncbi.nlm.nih.gov/books/NBK22448/.

[107] U. Schneider, P. Lunkenheimer, A. Pimenov, R. Brand, and A. Loidl. Wide range dielectric spectroscopy on glass-forming materials: An experimental overview. *Ferroelectrics*, 249(1):89–98, jan 2001. ISSN 0015-0193. doi:10.1080/00150190108214970.

[108] John G. Ferguson. Classification of Bridge Methods of Measuring Impedances*. *Bell System Technical Journal*, 12(4):452–468, oct 1933. ISSN 00058580. doi:10.1002/j.1538-7305.1933.tb00405.x.

[109] Daniel Rairigh, Andrew Mason, and Chao Yang. Analysis of On-Chip Impedance Spectroscopy Methodologies for Sensor Arrays. *Sensor Letters*, 4(4):398–402, dec 2006. ISSN 1546198X. doi:10.1166/sl.2006.054.

[110] Chao Yang, Daniel Rairigh, and Andrew Mason. Fully Integrated Impedance Spectroscopy Systems for Biochemical Sensor Array. In *2007 IEEE Biomedical Circuits and Systems Conference*, pages 21–24. IEEE, nov 2007. ISBN 978-1-4244-1524-3. doi:10.1109/BIOCAS.2007.4463299.

[111] Keysight Technologies. *Impedance Measurement Handbook: A guide to measurement technology and techniques*. Keysight Technologies, 6th edition, 2016. URL http://literature.cdn.keysight.com/litweb/pdf/5950-3000.pdf.

[112] A.K. Lu, G.W. Roberts, and D.A. Johns. A high-quality analog oscillator using oversampling D/A conversion techniques. *IEEE Transactions on Circuits and Systems II: Analog and Digital Signal Processing*, 41(7):437–444, jul 1994. ISSN 10577130. doi:10.1109/82.298375.

[113] Jorge Enrique Prada Rojas, Dr-Ing Wolfgang Krautschneider, and Dr-Ing Paola Vega Castillo. Design of a Wide Tuning-Range CMOS 130-nm Quadrature VCO for Cell

Impedance Spectroscopy. In IEEE, editor, *6th IEEE Germany Student Conference Proceedings*, pages 1–7, Hamburg, 2014. IEEE.

[114] Jörg Berkner and Klaus-Willi Pieper. Bondwire Inductance: Calculation and Measurement, 2014. URL https://www.iee.et.tu-dresden.de/iee/eb/forsch/AK-Bipo/2014/AKB2014_04_IFX_Berkner.pdf.

[115] Microwaves 101. Wirebond Impedance and Attenuation., 2018. URL https://www.microwaves101.com/encyclopedias/wirebond-impedance-and-attenuation.

[116] David M. Pozar. *Microwave Engineering*. John Wiley & Sons, 4th edition, 2011. ISBN 978-0-470-63155-3.

List of Figures

List of Tables

List of Abbreviations

ABBM Auto-Balancing Bridge Method

AC Alternating Current

ADC Analog-to-Digital Converter

ADS Keysight's Advanced Design System

ATP Adenosine Triphosphate

AMS Austria Microsystems

ASIC Application-Specific Integrated Circuit

BGA Ball Grid Array

BLC Blood Lactate Concentration

CE Counter Electrode

CPE Constant Phase Element

DC Direct Current

DDS Direct Digital Synthesis

DFT Discrete Fourier Transform

DI Deionized

DMEM Dulbecco's Modified Eagle Medium

DNA Deoxyribonucleic acid

DUT Device Under Test

ECIS Electric Cell-substrate Impedance Sensing

EIS Electrochemical Impedance Spectroscopy

EOC End Of Conversion

ERC Electric Rule Check

ESD Electrostatic Discharge

FC Flip-Chip Bonding

FFT Fast Fourier Transform

FPGA Field-Programmable Gate Array

FRA Frequency Response Analysis

GUI Graphical User Interface

HFSS High Frequency Structure Simulator

I2C Inter-Integrated Circuit

IHP Innovations for High Performance Microelectronics

IHP Inner Helmholtz Plane

ITCR Instituto Tecnológico de Costa Rica

LDH Lactate Dehydrogenase

LPF Low-Pass Filter

MLIN Microstrip Line

MCLIN Coupled Microstrip Line

OHP Outer Helmholtz Plane

PCB Printed Circuit Board

PSD Power Spectral Density

RE Reference Electrode

ROM Read-Only Memory

SAR Succesive-Approximation Register

SiGe Silicon Germanium

SIPO Serial-Input Parallel-Output

SLC Sweat Lactate Concentration

SMU Source Measurement Unit

SoC System on Chip

SOLT Short-Open-Loop-Through Callibration

SWR Standing Wave Ratio

TUHH Technische Universität Hamburg-Harburg

VNA Vector Network Analyzer

WB Wire Bonding

WE Working Electrode

WSE Working Sense Electrode

List of Symbols

A_e: Electrode surface area [m^2]

A_W: Warburg constant [$\Omega/\sqrt{s}$]

C_{dl}: Double-layer capacitance

C^r: Concentration of reductant species

C^o: Concentration of oxidant species

D: Diffusion coefficient [m^2/s]

f_c: Cut-off frequency [Hz]

f_{max}: Maximum oscillation frequency [Hz]

f_T: Cut-off frequency [Hz]

F: Faraday const. (96485.336 C/mol)

G: Geometric constant of electrodes

I: Current [A]

I_{CM}: Tail current of current mirrors

J: Current density [A/m^2]

k_B: Boltzmann const. (1.38×10^{-23} J/K)

M: Molar mass [mol/L]

N_A: Avogadro const. (6.02×10^{23} mol^{-1})

R: Ideal gas const. (8.314 J/mol·K)

R_{ct}: Charge-transfer resistance

r_{in}: Input resistance

R_f: Feedback resistance

R_s: Solution resistance

R^2: Coefficient of determination

SS: Sum of squares

t: Time [s]

T: Temperature [K]

v_i: Input voltage

v_o: Output voltage

V_{ac}: Alternating-current voltage

V_{cont}: Control voltage of varactors

V_{dc}: Direct-current voltage

V_{off}: Offset voltage

w_i: Weighting factor

z: Number of electrons transferred

Z: Impedance [Ω]

Z_0: Characteristic impedance [Ω]

Z_W: Warburg Impedance [Ω]

Γ: Reflection coefficient

ϵ: Permittivity [F/m]

ϵ_{c}: Permittivity of cells

ϵ_{m}: Permittivity of dielectric medium

ϵ_{mix}: Permittivity of dielectric mixture

ζ: Damping ratio

θ: Phase

ρ: Resistivity [$\Omega \cdot$ m]

σ: Conductivity [S/m]

τ: Time constant [s]

χ^2: Chi-square Goodness-of-Fit

ω: Angular frequency [rad/s]

Lebenslauf

Personal information

Name: Juan José Montero-Rodríguez

Birth date: April 7th, 1988

Nationality: Costa Rican

Studies

09/2014-08/2018	**Research Assistant** at the Institute of Nano- and Medical Electronics, Hamburg University of Technology (TUHH), Hamburg, Germany
02/2012-12/2013	**Master of Science** in **Microelectromechanical Systems**, Instituto Tecnológico de Costa Rica (ITCR), San José, Costa Rica
02/2007-12/2011	**Licentiate degree** in **Electronics Engineering**, Instituto Tecnológico de Costa Rica (ITCR), San José, Costa Rica

Working experience

01/2014-07/2014	**Lecturer** at the Electronics Engineering Program, Instituto Tecnológico de Costa Rica (ITCR), San José, Costa Rica
01/2012-05/2012	**Lecturer** at the Technical Program in Nanotechnology, Instituto Tecnológico de Costa Rica (ITCR), San José, Costa Rica

Bisher erschienene Bände der Reihe

Wissenschaftliche Beiträge zur Medizinelektronik

ISSN 2190-3905

1	Wjatscheslaw Galjan	Leistungseffiziente analogdigitale integrierte Schaltungen zur Aufnahme und Verarbeitung von biomedizinischen Signalen ISBN 978-3-8325-2475-3 37.50 EUR
2	Jakob M. Tomasik	Low-Noise and Low-Power CMOS Amplifiers, Analog Front-Ends, and Systems for Biomedical Signal Acquisition ISBN 978-3-8325-2481-4 37.00 EUR
3	Fabian Wagner	Automatisierte Partitionierung von Mixed-Signal Schaltungen für die Realisierung von Systems-in-Package ISBN 978-3-8325-3424-0 38.00 EUR
4	Mario A. Meza-Cuevas	Stimulation of Neurons by Electrical Means ISBN 978-3-8325-4152-1 45.00 EUR
5	Lait Abu Saleh	Implant System for the Recording of Internal Muscle Activity to Control a Hand Prosthesis ISBN 978-3-8325-4153-8 44.50 EUR
6	Andreas Bahr	Aufnahme von Hirnsignalen mit extrem miniaturisierten elektronischen Systemen zur Untersuchung von lokalen neuronalen Vernetzungen ISBN 978-3-8325-4525-3 43.50 EUR
7	Bibin John	Smart system for invasive measurement of biomedical parameters ISBN 978-3-8325-4536-9 44.50 EUR
8	Juan José Montero Rodríguez	Impedance spectroscopy for characterization of biological matter ISBN 978-3-8325-4746-2 48.00 EUR

Alle erschienenen Bücher können unter der angegebenen ISBN-Nummer direkt online (http://www.logos-verlag.de) oder per Fax (030 - 42 85 10 92) beim Logos Verlag Berlin bestellt werden.